European Military Museums

A Survey of Their Philosophy, Facilities, Programs, and Management

by J. Lee Westrate

SMITHSONIAN INSTITUTION
WASHINGTON
1961

U
13
.A1
W4
1961

SMITHSONIAN PUBLICATION 4432

FOREWORD

During the past century the Smithsonian Institution has published a number of studies relating to special aspects of the museum world, for the benefit of museum workers and those interested in the development of the philosophy and techniques of museology. The present volume reports the results of a study of certain selected European military museums made in 1958 by its author, Mr. J. Lee Westrate, for the President's Committee on the American Armed Forces Museum. Mr. Westrate served as the Committee's research director. His book contains a wealth of information that may prove of interest to those concerned with the general problems of organizing and maintaining museums. Although the study was restricted to military museums, the findings and the conclusions presented will be found to have a wider applicability to museums in general. Not only professional museum people, but also many laymen interested in the organized efforts of modern society to preserve important historic objects, will find the book informative and useful.

The President's Committee on the American Armed Forces Museum accepted Mr. Westrate's excellent report, essentially in its present form, and presented it to President Eisenhower. It may be pointed out, however, that the facts and opinions presented are exclusively those of the author and do not represent the opinions or attitudes either of individual members of the Committee or of the Smithsonian Institution.

The decades since World War II have seen a great development all over the world of museum buildings and the evolution of new and important museum techniques. It is hoped that the comparative observations, statistical data, and descriptive material in this book may stimulate similar comprehensive studies of other categories of specialized museums as they are currently being developed throughout the world. Readers are referred to Mr. Westrate's Preface for details of the scope and plan of the volume.

LEONARD CARMICHAEL
Secretary, Smithsonian Institution

CONTENTS

	Page
Foreword, by Leonard Carmichael	iii
Preface	viii

PART I: ESSENTIAL CHARACTERISTICS OF EUROPEAN MILITARY MUSEUMS

The Elements of Appraisal	1
Philosophy of the European Military Museum	5
Museum Buildings	12
Exhibition Techniques	19
Programs and Services	25
Management and Organization	28
Museum Finances	36

PART II: SIGNIFICANT MILITARY MUSEUMS AND WEAPONS COLLECTIONS

Imperial War Museum	39
National Maritime Museum	45
Tower Armouries	53
Royal United Service Museum	56
Musée Royal de l'Armée et d'Histoire Militaire	60
Leger en Wapenmuseum (Generaal Hoefer)	66
Musée de la Marine	72
Musée de l'Armée	83
Tøjhusmuseet	91
Haermuseet	99
Armémuseum	106
Statens Sjohistoriska Museum	117
Museum für Deutsche Geschichte	130
Heeresgeschichtliches Museum	141
Museo del Ejercito	152
Museo Naval	164
Museu Militar	171
Significant Special Arms Collections	177
Epilogue	201
Appendix: Space Allocations in European Military Museums	203
Index	205

PREFACE

In January 1958, President Dwight D. Eisenhower appointed a committee of distinguished public servants and private citizens to develop proposals for the establishment of an American armed forces museum. The President expressed his belief that such a museum could "make substantial contributions to our citizens' knowledge and understanding of American life." He invited Chief Justice Earl Warren to serve as chairman of the Committee and appointed as members Senator Clinton Anderson, Senator Leverett Saltonstall, Senator H. Alexander Smith, Representative Clarence Cannon, Representative Overton Brooks, Representative John Vorys, Secretary of Defense Neil McElroy, Governor Nelson A. Rockefeller, Gen. Kenyon A. Joyce, Dr. John Nicholas Brown, and Dr. Leonard Carmichael, Secretary of the Smithsonian Institution. The President also requested Dr. Carmichael to serve as the Committee's executive director.

In undertaking its significant responsibility of acquiring pertinent facts upon which to make soundly based decisions, the Committee recognized that Europe is a very productive source of useful information on the subject of military museums. In establishing and developing their own military museums, European nations have dealt with many matters which confronted the President's Committee. In some instances, their experience in this field dates back over a century, but no recent comprehensive analysis has been published. The Committee concluded that a study of this nature would be most profitable in the anticipation that the experiences of European military museums would provide a source of valuable insight into the many facets of such activity.

Upon instruction of the Committee, its research director selected a number of well-known and highly regarded European military museums for study. During the summer months of 1958, he visited more than 25 museums in 13 countries and made a detailed examination of their exhibits, administration, and other essential operations. Many of these museums are national in scope and graphically present materials which deal with an extended period of military history. Few give coverage to all the armed services; the normal pattern is a museum for either the country's army or navy. Although they contain interesting and excellent collections, regimental and unit museums were not studied. The information they could provide would have had little application to the problems confronting the President's Committee. The military exhibits contained in some of the larger general museums of Europe were also visited because their scope of coverage is often similar to that of the military museum. However, they must compete with other museum departments for space and resources and are thus somewhat restricted in their programs and exhibits. This study of course could not comprehend all European military museums.

The principal data for this study were obtained from the Imperial War Museum, the Royal United Service Museum, and the Tower Armouries in London; the National Maritime Museum at Greenwich; Le Musée Royal de l'Armée et d'Histoire Militaire in Brussels; the Leger en Wapenmuseum (Generaal Hoefer) at Leiden; the Tøjhusmuseet in Copenhagen; the Haermuseet and the Oldsaksamling at the University of Oslo; the Armémuseum and the Sjohistoriska Museum in Stockholm; the Heeresgeschichtliches Museum and Kunsthistorisches Museum in Vienna; the Musée de l'Armée and the Musée de la Marine in Paris; the Museo del Ejercito and the Museo Naval in Madrid; and the Museu Militar in Lisbon. Other useful information was obtained from departmental curators at the Museum für Hamburgische Geschichte in Hamburg, the Bayerisches Nationalmuseum in Munich, the Schweizer Landesmuseum in Zurich, the Castel San Angelo in Rome, and the Real Armería in Madrid. A visit to the Museum für Deutsche Geschichte in East Berlin permitted first-hand observation of a military museum in a Communist-dominated nation and provided a basis for comparing the role of such a museum with that of its counterparts in democratic countries.

In presenting this material, two distinct but complementary methods have been used. The first few sections examine the essential characteristics of European military museums within the framework of certain major generalizations. Military museums normally fulfill the same basic functions, engage in identical or similar activities, and are confronted with the same problems regardless of location. These can be discussed in general terms, but the student of military museums finds that some detailed information about each is of equal interest. Individual museums achieve similar objectives by different methods, view their responsibilities in a different manner, do not enjoy the same assets or suffer from the same limitations, possess professional staffs of varying degrees of competence, make widely varied contributions to the study of military history, and achieve a high level of technical expertise in different fields of museum specialization. This type information can best be presented in a brief description of an individual museum. Hence, the concluding sections of this study have been written in the form of sketches for some of the leading military museums and significant weapons collections in Europe.

The preparation of this volume would have been impossible without the splendid cooperation provided by European museum directors and their staffs. They willingly furnished photographs, some of which are herein reproduced, and engaged in candid and detailed discussion about their operations, activities, accomplishments, limitations, and problems. They volunteered their advice and offered a number of admonitions based upon personal experience which they thought might be useful for an embryonic institution. They all expressed their hope that a new national armed forces museum might some day be founded in the United States and indicated their eagerness to be of assistance where possible. United States Embassy officials further simplified the problem of research by their excellent liaison with local museums. Grateful appreciation must also be expressed to the members of the President's Committee on the American Armed Forces Museum whose quest for vital information on the subject of military museums caused this study to be undertaken.

The author wishes to pay particular tribute to Maj. Gen. Kenyon A. Joyce, who played a significant role in the work of the President's Committee on the American Armed Forces Museum. He participated in the preliminary study that preceded creation of the Committee and served as chairman of the subcommittee that guided the Committee's research activities. His untimely death before the Committee completed its labors deprived his colleagues of his wise counsel and untiring interest born of a distinguished career in the service of his country.

J. Lee Westrate

Part I: Essential Characteristics of European Military Museums

THE ELEMENTS OF APPRAISAL

The curators of military museums in the United States have long been aware of the extensive experience their European colleagues have had in this specialized field of museology. A number have visited the many fine military museums of Europe to examine and study their extensive collections of historic and treasured arms and armor and to exchange professional information on a variety of pertinent subjects. However, there has been little inclination to study European military museums as institutions apart from a general awareness of their facilities and some of the specialized services they perform. Even as these museums have confronted and found solutions to the problems of restoration and preservation of specimens, modes of exhibition, and documentation of artifacts, they have also gained a wealth of experience in coping with the varied problems of management. While it is true that European museums have been fashioned within the military traditions they seek to portray and as a result they differ in many substantive aspects from existing American military museums or any which might emerge in the future, they are able to contribute their vast knowledge about the many facets of museum activity they share in common.

Regardless of its size, each national European military museum is a somewhat complex institution of absorbing interest. The student of military museums can spend countless hours poring over the contents of weapons collections or make an exhaustive analysis of the many details of museum operations. The compilation of such data would be a valuable contribution to an ever-increasing store of museological information. This study, however, was not undertaken with the objective of such intense examination. Rather, an attempt was made to cull from a vast body of information those facts which appear significant in fashioning the museum as a particular type of institution. The minutia of extraneous, though often interesting, detail has been avoided to focus upon the principal aspects of museum operations and activity. Hence, six key subjects have been selected for some detailed consideration:

Museum Philosophy. Every military museum presumably has a fundamental philosophy which is given expression in some discernible objectives. This philosophy is usually apparent in the exhibits and the type of services which the museum makes available to the general public, although its more detailed aspects may be discovered only after prolonged discussion with its staff. It is possible to evaluate in part a museum's effectiveness by noting its stated objectives, by judging if its resources are commensurate with them, and by observing the manner in which these objectives are implemented in its exhibits, programs, and services.

Building Facilities. The building in which the military museum is housed can create major problems for the museum staff. If the structure occupied was originally not built as a museum and is of considerable antiquity or somewhat inadequate for use as a museum, it imposes many severe limitations upon the kind of exhibits which can be shown to the public or creates insurmountable obstacles with which the museum staff is virtually unable to cope. In this analysis of European museums, the type and condition of the buildings were studied and related to the type and quantity of specimens displayed. This was attempted in an effort to formulate some conclusions regarding the determination of adequate space for exhibition, ready reference, storage, offices, workshops, and other facilities which are required for a military museum.

Exhibition Techniques. The display of military artifacts requires the application of specialized techniques. These are contingent upon the size and quantity of weapons, armor, uniforms, and other objects in the museum collections; space available; the amount of money which can be spent on building the exhibits; the philosophy of the museum; and the competence and ingenuity of the museum staff. An effort was made to discover those display techniques employed in Europe which might prove particularly useful to other museologists. The exhibits in the European military museums visited were evaluated as to their over-all esthetic appearance, the manner in which particular objects were displayed, the utilization of space available for exhibition, and the considerations extended to the viewer, insofar as they might be discovered through casual observation. The condition and preservation of individual specimens are also of particular interest, and the type of expert services which are performed by individual museums, such as the restoration of colors, prints, and paintings, was duly catalogued.

Museum Programs and Public Services. In part, these vital areas of activity are related to museum philosophy and capabilities. The objectives of some European military museums are quite limited or focused and the staff is correspondingly small. Other museums plan and execute a broad range of activities and services for the public with a correspondingly larger staff. These latter institutions are normally in much sounder financial condition than those which are considerably limited in what they can do. Regardless of the size and scope of any particular museum, there are a number of services every museum is expected to provide the public. However, the extent to which these are furnished also relates principally to the factors of philosophy and staff competence.

Management and Organization. It is axiomatic that the quality of administration determines the ultimate efficacy of an organization, regardless of the resources at its disposal. In closely studying European military museums of considerable experience, there was the natural inclination to search for the ultimate model of administration, but this model was not found. However, an administrative analysis of these museums required that answers be sought to several basic questions. These are: Does the organizational structure appear to be a logical one? What policy-making body exists, what authority does it exercise, and what policy and administrative discretion is accorded to the chief executive officer? Are the lines of control clear from the museum director through the department heads to rank-and-file employees? Is there evidence of coordination among the various divisions or departments of the museum, so that the basic philosophy of the museum is given expression in each? Is adequate recognition given to the professional people and technicians on the staff, and are there sufficient incentives for each individual to strive for a high degree of personal accomplishment? Finally, is the organization administered in such a way that human and other resources are utilized with maximum efficiency?

Finance. The amount of money available to operate a military museum is certainly an important factor in ultimately determining its scope and quality. Experience has proved that museums with strong financial support are those which achieve high quality in their exhibits, are able to employ a staff of superior competence, maintain an ambitious program of scholarly endeavor, and can provide the public with many useful services. Of primary interest are the principal sources from which European military museums receive their income, the distribution of major items within the museum budget, and the provisions made for building improvements and upkeep.

In some respects these subjects of inquiry are but the supporting elements necessary to provide an apt setting for the display of the museum's treasures. They are the elements that are summoned into consideration in evaluating the significance, contribution, and general excellence of a particular military museum, or any type of museum for that matter. Yet, most often the public is attracted to the museum because of the quality, value, and celebrated nature of the objects on display, not by the broad base of its philosophy, the architectural splendor of its building, the artistic quality of its exhibits, the scope of its program, the soundness of its financial position, or because it is a model of administrative efficiency. The objects are what people come to see, and European military museums are truly well endowed with the highly prized memorabilia of warfare. Nevertheless, these important aspects of museum operations have a profound impact upon the type of experience the visitor has when he comes to the museum.

PHILOSOPHY OF THE EUROPEAN MILITARY MUSEUM

A museum performs a unique role in recording man's history, for it has the responsibility of systematically collecting, preserving, and exhibiting significant objects which are historically important to the development of a particular society or culture. Whether its collections cover many fields of knowledge or are limited to one specialized area, the role is the same: to provide through the subject matter of its exhibits a graphic presentation of some period of history. In this respect the role of the military museum is not dissimilar from that of any other museum. Its general subject area is warfare and military science, and its artifacts are the weapons, techniques, and equipment employed by the serviceman as he experienced the incidents of military history.

In conjecturing about the philosophy of any particular military museum, it is necessary to observe that the museum has four basic functions. This philosophy can often be discovered by noting the emphasis it gives to each of these and the methods it adopts in giving expression to its objectives. It is doubtful that any museum can ascribe equal weight to each, although these functions are not mutually antagonistic. However, the tendency is to emphasize one or two at the expense of the others. The choice of which function is to take precedence is not often a free one. It may be, and often is, determined by the amount of money available to the museum for its total operations. A museum director may also have his course definitively charted by the museum's authorizing legislation, the views of its policy-making body, or the limited competence of his staff. Nevertheless, each museum attempts to carry out in some form its four basic functions.

Custodial. Every museum is a repository of specimens. Each brings together objects of historical importance and assures their preservation for posterity. The museum performs the useful service of systematic collection, identification, and description, and relates these objects to the particular period of history in which they may have been used. The final and ultimate responsibility of the museum as custodian is to make the specimens available to public view and study. In doing so it must choose whether to display only those significant artifacts which will make a real contribution to the value of its exhibition, or to show the public its entire collection. There is a great temptation to do the latter, for military museums often acquire extensive collections of the same type of weapon, with each item in the collection having only slight variations from the others. The museum staff may feel that the public will be properly impressed by the museum's inventory only if it shows all or most of such items in its possession. Often this is the convenient way to solve the problem of storage, for many museums have a large number of objects and very limited space in which to place them. It is logical to assume that the custodial function is the one emphasized in most military museums, for the viewer enters expecting to see weapons and the other paraphernalia of warfare. He is usually not disappointed. Often he is inundated and gets the impression he has come into an armory.

Educational. Regardless of the form its exhibition takes, the military museum is attempting to tell a story. The story may be, and often is, very simple. A collection of arms used during a particular war may be displayed without any references to the history of the war itself. The same may be done for uniforms and insignia. A visitor to the museum might be able to discover the evolution of the cannon or other types of artillery by examining a display which has been arranged chronologically. In this manner he is able to achieve a familiarity with objects somewhat apart from any context of military history. This is the simplest educative technique and about the extent to which it is attempted by some very impressive museums.

Experience has shown that museums must exercise some restraint in fulfilling their educational responsibilities to the general public. All semblance of deliberateness must be avoided, for the average visitor develops a certain amount of natural resistance when he detects that an obvious attempt is being made to instruct him. The museum must employ considerable subtlety in presenting the information it desires to convey to the visitor. This is largely accomplished through attractive displays which present ideas through tested visual educational techniques, accompanied by explanatory material which is both brief and concise. Dr. Walter Havernick, director of the Museum für Hamburgische Geschichte in Hamburg, Germany, spoke of this to his colleagues attending the First Congress of Museums of Arms and Military Equipment at Copenhagen in May 1957:

In spite of all our efforts with presentations and displays in the museums . . . we have to be careful to refrain from too much systematics and pedagogies if we wish to get people into the museums. We must remember . . . that the visit to the museum should be an adventure, and nobody will go where they know that attempts will be made to educate them in a schoolmasterly manner. People will not go to a museum on a Sunday forenoon to be educated.

The museum also performs an important educational function for the specialist. It gives him access to the collections and lets him pursue his own quest for specialized knowledge. Members of the museum staff may assist such a student or may be pursuing their own independent course of research. If the museum takes its educative functions seriously, it may provide guided tours by its professional staff, conduct special lectures and seminars, and make its facilities available for the meetings of groups interested in the field of military history.

Commemorative. Military museums have a natural tendency to memorialize heroes and organizations. A national military museum usually has been established to serve as a suitable memorial to a particular service or all the armed forces represented within its collections. The memory of a great leader is perpetuated because the museum has acquired his uniforms, personal weapons, medals, and other souvenirs and because any epoch in military history cannot be adequately accounted for without reference to important leaders. Many national European military museums tend somewhat to avoid an over-emphasis on military leaders, but they recognize the usefulness of memorialization in attracting the public. However, some frankly build their exhibits around the exploits of national heroes and sovereigns.

Entertainment. Apart from the serious student of military specimens or the rather limited group with a keen interest in such things, most people come to a military museum because they expect to see something interesting. In short, they expect to be entertained and carry away the memory of a pleasant experience. If the museum staff is deeply immersed in its professional and academic interests, it may forget that not many of the public are so absorbed and fail to make the museum a source of real entertainment. Such a possibility was suggested by Commander William E. May, deputy director of the National Maritime Museum in Greenwich in a discussion before the Copenhagen

Congress of Museums of Arms and Military Equipment. Commander May said he believed museum directors "often feel that the presence of the public is an intrusion upon our chosen interests." He then recalled a frequent statement by the original director of the National Maritime Museum that "a museum would be heaven if it were not for the need to have visitors in it." While museum staffs may develop a certain distaste for pandering to the viewer, they must take him into consideration and make their exhibits attractive if their work is to be undergirded by strong public support and other objectives of the museum are to be achieved. In the mind of many a viewer, the excellence of the museum is determined by its physical attractiveness rather than by the quality of its collections. The viewer is more likely to get an over-all impression of the museum (possibly an unfavorable one) and overlook some excellent items unless his attention is drawn to them by an attractive display.

The military museums of Europe run the full gamut of objectives. A few try to portray graphically the full impact of modern war upon their nation, whereas others seek only to provide the public with a display of arms and armor. In between these two extremes are the museums which depict some aspects of military history through the display of pertinent weapons, uniforms, equipment, and pictures. Most museums are not particularly interested in showing peacetime contributions of the armed services, although they often display weapons and uniforms used by the armed services in time of peace.

The most ambitious program and expanded role of a European military museum as visualized by its staff appears to be that of the Imperial War Museum in London. In its published brochure, the museum modestly states that it illustrates and records all aspects of the World Wars of 1914-1918 and 1939-1945 and the other operations in which Forces of the British Commonwealth have been engaged since August 1914, and makes accessible to the public and to students data of all kinds relating to them. Thus, it places greatest emphasis upon its custodial and educational functions and views its major contribution as providing a center for intensive study of modern British military history. Unfortunately, its total resources are not sufficient to achieve the full potential of its ambitious objectives. What is important, however, is that the Imperial War Museum sees its role as a dynamic one in an age where military power has a tremendous impact upon national and international institutions.

The philosophy of Le Musée Royal de l'Armée et d'Histoire Militaire in Brussels contrasts somewhat with that of the Imperial War Museum, although the two museums display the same general range of artifacts. The Brussels Museum appears to have expended much effort assembling a great quantity of worthwhile Belgian and foreign military objects. This museum stresses its custodial function above all others, for it displays nearly all the specimens in its collections. Although the museum has an excellent library, it gives considerably less attention to original research than does the Imperial War Museum. However, the impression should not be gained that the Brussels Museum neglects the function of educating the public about Belgian military history. The exhibits in chronological halls are so crammed with objects that the careful and tenacious viewer learns a great deal about the subject whether or not he had that intention when he entered.

The Army History (Heeresgeschichtliches) Museum in Vienna appears to have a philosophy which lies somewhere between that of the Imperial War Museum and the Army Museum in Brussels. Its professional staff is composed of military historians who also serve as part-time faculty members at the local university. For this reason, the museum gives a strong academic emphasis to its treatment of military history, and much scholarly endeavor originates at the museum, both through the personal initiative of the staff and at the request of the university. In its custodial function, the Vienna Museum concentrates upon the display of significant artifacts in an artistic setting, thus assuring a strong flavor of the fine arts in its exhibits. Because of this attention to esthetic matters, the Heeresgeschichtliches Museum blends into the pattern of general grandeur often associated with Viennese museums. This tends to increase its value as a source of entertainment for the viewer. The museum's commemorative aspect is very strongly emphasized in Vienna, and much attention is given to the display in separate halls of objects connected with Prince Eugen, Empress Maria Theresa, Archduke Charles, and the famous Austrian Commander Radetzky. In this military museum perhaps one finds a closer balance of the four museum functions than appears to be achieved anywhere else in Europe.

The Musée de l'Armée in Paris places about equal emphasis upon the custodial and commemorative functions. It possesses an extensive and excellent collection of swords, armor, and guns, which cover the full range of French Army history and displays them in large quantities in a series of topical halls. However, the museum's principal attraction is its large collection of Napoleonic relics, and in a very real sense the museum is a memorial to Napoleon Bonaparte, the most celebrated of all French military heroes. Indeed, it would be most difficult for the museum to avoid concentrating on the Emperor's career, for the Musée de l'Armée is located at the Hôtel des Invalides, the site of Napoleon's tomb.

The three largest naval museums in Europe appear to have borrowed heavily from one another both in objectives and methods of display. The National Maritime Museum in Greenwich, England, the Musée de la Marine in Paris, and the National Maritime (Sjohistoriska) Museum in Stockholm announce a similar purpose in their brochures. They all exist to commemorate the exploits of their country's sea power and to provide the student with some comprehensive coverage of the nation's naval history. Each also provides extensive exhibits on the merchant navy, fishing fleet, and pleasure craft. Each emphasizes the chronological evolution of naval power and limits its auxiliary research program to the objects on display and a few special projects which absorb the interests of individual staff members during their spare time. The similarity of their exhibits is striking. Because each museum has a large collection of excellent ship models, the interest of the viewer is held for a lengthy period. The museum staffs also appear to have recognized the value of entertaining the visitor through the use of extremely attractive display techniques. Each museum is very impressive and leaves a favorable impact upon the visitor. By thus stimulating the viewer's interest, these museums achieve in part their stated objectives.

The very excellent Army Museum in Stockholm follows a pattern similar to these three naval museums in its presentation of the sweep of Swedish Army history. However, its research program is much stronger, largely owing to the leadership of the museum director who ranks as one of the leading military historians in Europe.

The arsenal type museums have the most restricted objectives. They exist primarily for one purpose—to display weapons and armor. They may also exhibit some uniforms, insignia, and medals, but these are secondary to the weapons themselves. No effort is usually made to commemorate military heroes, but this may occur if the museum acquires a personal collection of a leading military figure and displays it as a unit. The arsenal museums often have an intensive research program related to their collections. These studies tend to be specialized treatises on the weapons themselves or closely related subjects. Such museums are also a haven for the specialist, for the collections are often sufficiently extensive to enable an individual to develop a high level of expertise through concentrated study. The appeal of such museums to the public tends to be limited, and considerable ingenuity in display techniques is required if large quantities of arms are to be displayed attractively, without boring the average viewer. However, a good collection of armor is an exception to this. Excellent pieces are examples of fine art and can be attractively shown to the public.

The Tøjhusmuseet in Copenhagen is perhaps the outstanding arsenal museum in Europe. The Leger en Wapenmuseum in Leiden is in process of completing its work of restoration and is similar to the Tøjhusmuseet, although considerably more limited in scope. The Tower of London Armouries, the Real Armería in Madrid, the Waffensammlung in the Kunsthistorisches Museum of Vienna, and the Department of Arms in the Bavarian National Museum of Munich are the best examples of a combination armor and weapons collections of particular periods. In each case the objective is simple—the display of arms and armor.

None of the European museums embrace the full philosophy that would probably characterize a national armed forces museum in the United States. The Imperial War Museum suggests the possible broad scope of such an institution, and the Army Museum in Vienna affords a good example of a museum which has achieved a commendable balance of museum functions. The rest are somewhat restricted in purpose but offer useful suggestions of practical objectives which can be considered within specialized areas.

MUSEUM BUILDINGS

It is necessary to stress that museum housing presents major difficulties to most of the military curators in Europe. With an occasional exception, the buildings now used as military museums were not originally constructed for that purpose, and it has been the usual lot of the military museum to inherit whatever structure was available at the time of its founding. Museum staffs have been confronted with the great task of adapting such a building to its new purposes. They have often struggled against insurmountable obstacles. The results of their labors have been quite unsatisfactory in many instances, but they cannot be blamed for many of the failures. In some cases the buildings are of such a design that even with very extensive renovation it would still be quite impossible to adjust exhibit space to the requirements of the best museum techniques or to plan a coordinated set of galleries. Almost all military museum buildings are extremely old and possess many of the faults which modern architects have tried to correct. Thus, there is often a built-in resistance to the installation of modern lighting techniques, suitable workshops, offices, and other facilities vital to museum operations.

The almost universal cry of the military museum director is that he has insufficient space. His first complaint is that he does not have enough exhibition space, but even if the building is deemed large enough, the size and configuration of particular rooms raise pressing problems. Another complaint is that storage space is inadequate. If the museum director decides he will display only a few significant artifacts rather than whole collections, the problem of storage becomes very acute. The inclination has been to follow the path of least resistance and place as many objects as possible on display, thus eliminating the need for a large amount of storage space. A museum is indeed fortunate if it has a separate building which can be used for storage; usually storage area can be gained only at the expense of exhibit space or by decreasing the room allocated for other facilities.

Related to exhibit and storage space problems is one of finding room for housing reference collections. To be of any practical use, reference collections must be readily accessible to the public or to the serious student. Hence, the problem of housing is considerably more technical for the reference collection than for what might be called dead or regular storage. The problem of locating the items in ready reference has also been solved in several ways. Some museums have placed reference collections in the same room with their prototype exhibits. Others have been able to house the collections in separate rooms. Still other museums have eliminated reference collections almost entirely or reduced the number to only a few significant ones.

Finally, there is the complaint of inadequate office space or insufficient room for workshops and employee facilities. The museum director is again confronted with the problem of how best to allocate existing space so that the irreducible needs of the museum are met.

The difficulties caused by physical plant can be seen most readily in the type of buildings which now house European military collections. Only a few may be discussed here, but they illustrate the problem quite thoroughly. The Imperial War Museum in London has one of the most complicated housing problems, for it occupies a building which had originally been erected in the early 1800's as a hospital. When the structure was acquired, the museum staff had to face the prospect of placing military collections in long corridors and small cubicles. There is a ready admission that the problem resulting has not been solved satisfactorily, but that the best must be made of a very bad situation.

Two other military museums in London are in considerably better shape than the Imperial War Museum. The Royal United Service Museum is located in the Banqueting House of the original Whitehall Palace. This is a moderately spacious building which readily lends itself to the housing of military objects. The National Maritime Museum in Greenwich is located in three buildings. Two long, rectangular structures provide excellent opportunities for employing modern museum techniques. Between the two buildings is the historic Queen's House. Although this famous house might be used for military exhibits, the museum staff prefers to keep the building free from such encumbrances and thus more in the character of its original use as a royal residence and show place. The Armouries at the Tower of London, the other major military museum in the city, are housed in the oldest building in London and provide an interesting study in all the problems of space which have been mentioned, with the possible exception of exhibition area.

The location of the military museum in an ancient structure is quite common throughout the rest of northern Europe. For example, the weapons collection in Leiden is located in a building now approximately 250 years old. While there has been some renovation to house military objects, there are still certain problems of dinginess, difficult lighting, cleanliness, and other matters with which the staff has been able to deal in little more than a superficial manner. The Tøjhusmuseet in Copenhagen is located in an arsenal 350 years old. The museum's staff has capitalized on the building's original use and has attempted to display its specimens in such a way as to create the impression that the structure still is an arsenal. This museum is also fortunate in having separate buildings for storage and offices. The storage building, which is almost 350 years old and originally contained the King's Brewery, still requires a great deal of renovation before it can function adequately. The Army Museum in Oslo has received three of the outbuildings of Akershus Castle for its exhibits and storage space. These buildings are about 250 years old and have required considerable remodeling to house military collections. The Army Museum in Stockholm is somewhat similar to the Tøjhusmuseet in Copenhagen, for its building was originally the city's main artillery depot.

The inclination to use old government buildings that are obsolete for other purposes also exists in Madrid and Lisbon. In both cities, museums have been established in buildings which the government has abandoned for other purposes. These structures present the museum staffs with problems identical to those confronted by their colleagues elsewhere. They apparently have had some remodeling to meet the minimum needs, but they are still in need of much improvement. The Navy Museum in Madrid is located in the same building with the Ministry of the Navy. The results provide the ready conclusion that an office building is little more acceptable for a modern museum than some of the other types which have been discussed.

The use of an ancient structure for a military museum is best illustrated by the location of a military collection in the Castel San Angelo in Rome. The foundations of this building are alleged to have been laid over 2,000 years ago, and over the centuries it has been used as a tomb for a Roman emperor, as a Roman fortress, and as the residence and offices of the Pope. The present top structure of the building was erected during the medieval period. Although the Castel San Angelo is one of the most famous buildings in Rome and provides the visitor with splendid examples of ancient Roman architecture and medieval construction methods, it is hardly the place where one can expect to find modern museum techniques employed to display military objects. Nevertheless, one of its key attractions is a somewhat extensive collection of 15th- and 16th-century weapons.

There are some relatively new museum buildings in Europe. Although the structure housing the Army Museum in Brussels was not originally built for that purpose, it was erected about 50 years ago for an international exposition, and hence is very suitable for the display of military objects. It is also one of the largest military museums in Europe. The Navy Museum in Paris is located at the Palais de Chaillot which is a fairly modern structure. The Navy Museum in Stockholm was built in the 1930's and has curved halls similar to those of the Palais de Chaillot. Here is one example in which a European museum now housing a military collection was actually designed for this purpose.

The Army Museum in Vienna is a special case. The original building was severely damaged during World War II; all that remained at the end of the war was the central foyer. New halls were built, but they were constructed in keeping with the architecture of the cupola-crowned foyer and original building. In discussing the restoration of the museum, a key member of the staff stated that the wartime destruction proved very beneficial, for most of the defects existing in the original museum were corrected in the rebuilding. The problem of space was solved very satisfactorily. The service departments and storage area were all placed in one adjacent building and the offices in another so that the museum now features one building devoted entirely to display. The artillery collections are also located apart from the exhibit building. These are housed within colonnades which connect the office building and the workshops on one side and the building which some day will house the World War II collection on the other side. Thus, the museum in Vienna has been most fortunate in being able to eliminate many of its building problems through reconstruction.

Examination of the military museums in Europe readily confirms the complaints about inadequate space. However, it is necessary to look behind these complaints and attempt to determine if existing space is being used efficiently. Part of the answer is found in the type and quantity of specimens the museum has to display and the museum director's decision regarding the kind of exhibits he wishes to build. The problem is most acute where the museum has a very large inventory and a relatively small building. In such a situation, there is a very real temptation to place as many specimens as possible into the exhibits.

Where crowding occurs, the quality of the display correspondingly suffers. If the museum director resists the temptation to display quantity and decides that the viewer should see only striking objects, he often creates quite a problem of storage for himself. He must place a large number of items in dead storage, reduce the room allocated for offices, work space, and ready reference in order to find more area for storage, or eliminate as many items as possible from his inventory. Regardless of how the museum director may solve his space problem, the ultimate question of efficient space ultilization is left to his own ingenuity. Various techniques are applied with some very interesting and often pleasing results. It cannot be denied, however, that there is really no substitute for additional space.

Many museums give the immediate impression they are too crowded and that a much larger building is needed if the exhibits are to be attractive. The Imperial War Museum in London is a classic example of a museum which appears to need much more space. The space available for storage is quite small when compared to that allocated for exhibits and ready reference materials. The exhibition halls are crammed with objects and the viewer requires considerable tenacity if he looks at many of them. Many visits are suggested if the contents are to be seen thoroughly. The Army Museum in Brussels has the largest amount of exhibit space of any military museum in western Europe. However, it too is very crowded and gives the impression that almost every possible bit of floor and wall area is used to display some object.

A similar situation exists in the Museu Militar in Lisbon and the Naval and Army museums in Madrid.

Several museums appear to make fairly efficient use of their building space. For example, the National Maritime Museum in Greenwich, England, has a very large exhibition area and displays objects relatively small in size. Thus, there are few problems connected with showing large-sized items. There is likewise no feeling of crowding. Other museums which have accomplished an equally effective use of space are the Tøjhusmuseet in Copenhagen, the Army Museum in Stockholm, the Navy museums in Paris and Stockholm, and the Army Museum in Vienna. The two Naval museums seemingly display a great number of specimens. However, there is very little evidence of crowding. The displays of the Tøjhusmuseet are not too crowded at present. When the storage building is renovated, the number of objects currently on exhibit can be reduced. With the possible exception of the National Maritime Museum in Greenwich, the Army Museum in Vienna appears to have the least space problems. Floor and exterior space allocations of a number of European museums are described in the Appendix (p. 203).

Unfortunately, no definite standards have been established for determining precisely what the proper division of space within a museum should be. Basically such determinations are a matter of conjecture and are based upon several criteria of judgment. In most cases decisions as to the amount of space allocated for exhibits, workshops, offices, and storage are made as a result of trial and error. Exhibit space is primarily determined by the size, shape, and quantity of objects and by deciding how much to show. The size of the workshops is determined by the type of work done by the museum and the kind of equipment needed. Some museums do everything from restoring objects to building showcases. Others operate on a much more limited scale, and workshops are correspondingly smaller. The final decision regarding the division of space is primarily one of achieving a balance, so that all museum functions can be performed with an acceptable standard of adequacy.

The problem of what to do with an ancient building not susceptible to renovation is beyond the competence of the museum director. It is something he must live with. European directors indicate this is one of their greatest frustrations, but they have been able to accommodate to the realities of their problem and proceed from there.

The location of the museum is also of interest. Most European museums are well located either in the heart of the city or adjacent to the central tourist area. Those museums so fortunately situated are likely to have a fairly large number of visitors. Museums which are somewhat removed from the principal tourist area normally suffer a corresponding cut in the number of people who come to view the exhibits. Further, the clientele of a military museum is a somewhat select one. Often only those genuinely interested will search for a collection that is not within the tourist area, unless the museum is sufficiently impressive to attract the more inveterate sight-seer.

Examples of museums that are located in central tourist centers include the Tower of London, the Royal United Service Museum in London, the Tøjhusmuseet in Copenhagen, the Army Museum in Stockholm, the Musée de l'Armée and the Musée de la Marine in Paris, the military collection housed in the Castel San Angelo in Rome, and the Army and Navy museums in Madrid. Several military museums are somewhat outside the principal tourist area but still quite accessible to the public. These include the Imperial War Museum in London, the Leger en Wapenmuseum in Leiden, the Naval Museum in Stockholm, and the Army Museum in Vienna. The National Maritime Museum in Greenwich is the only major military museum in England outside the heart of London. However, it is located at the site of the Royal Naval College and the Greenwich Observatory and is in the center of a special tourist area. The Army Museum in Brussels is also something of a unique case. It too is somewhat outside the regular tourist center of Brussels. However, it is located at the Palais du Cinquantenaire, which is itself a tourist attraction. Thus, it is able to draw some visitors from among those who come to see the Cinquantenaire Arch and a neighboring museum. The following list shows the approximate number of visitors during 1957 for some principal European military museums:

Museums	Number of Visitors
Imperial War Museum, London	245,000
National Maritime Museum, Greenwich	400,000
Musée Royal de l'Armée, Brussels	120,000
Leger en Wapenmuseum, Leiden	11,000
Tøjhusmuseet, Copenhagen	80,000
Musée de la Marine, Paris	150,000
Armémuseum, Stockholm	25,000
Sjohistoriska Museum, Stockholm	40,000
Heeresgeschichtliches Museum, Vienna	400,000
Museo del Ejercito, Madrid	190,000
Museo Naval, Madrid	30,000
Museu Militar, Lisbon	14,000

EXHIBITION TECHNIQUES

Stress has been given to the impact which the size and condition of museum buildings produce upon the type of exhibits they will contain. To these factors must be added the ingenuity for constructing displays and the attitudes regarding preparation of exhibits which prevail among museum staff members. Many museum directors and individual curators pointedly explain the state of their exhibits in terms of difficulties produced by an inadequate building or other deficiencies largely beyond their control. Also they often express the opinion that Europe can demonstrate very little to the United States in the way of new display techniques and that the United States has now emerged to the forefront of this field. Quite often their exhibits confirm their views.

The experience of the European military museum proves conclusively that superb exhibits cannot be placed in buildings which fail to meet the minimums of adequate museum design without overcoming almost insurmountable difficulties. Also, they cannot be built to best advantage without a reasonable amount of funds. A building that is in poor condition, in need of extensive renovation, encrusted with age, poorly lighted, and largely precluded from other than superficial changes can dishearten the most energetic curator. At best, he may be able to produce an exhibit which is a makeshift compromise with his desires. Some museums have developed a long-range program of display improvement, but the success of this is dependent upon having the staff and technicians as well as the necessary funds to do it.

It is impossible to indulge in sweeping generalities about the quality and character of European exhibit techniques. It should be emphasized that most military curators of Europe have a high degree of professionalization, but some readily admit to grave limitations in exhibit preparation. However, many state they are looking for new techniques with which they can experiment. The impression is gained that a sincere effort has been made in most museums to do a good job, although exhibit experts might be able to suggest how additional improvement could be obtained by applying some inexpensive techniques.

European military museums provide examples of displays which touch about every degree of excellence. The viewer's personal reaction toward what he sees is a matter of individual judgment, but there appears to be a fairly broad area of agreement on what constitutes an attractive, pleasing, and appealing exhibit. Two distinct impressions of the exhibits in European military museums invariably emerge during a visit. The first results when entire halls are looked at in perspective. Thus, an over-all reaction is obtained. The second is quite restricted and is formed about individual exhibits or showcases. It is useful to keep the two separated, for delightful and attractive individual displays can often be found in halls that are crowded and poorly arranged.

The general subject of public reaction to the military museum came in for considerable discussion at the 1957 Congress of Museums of Arms and Military Equipment at Copenhagen. Dr. Heribert Seitz, director of the Armémuseum in Stockholm, insisted that the general public is prone to evaluate the whole museum in terms of personal reaction to the exhibits, that "from the visual angle of the public the work of the museum is, in spite of everything, most often judged by the way in which the exhibition is arranged—because it is easier for a layman to judge the outer shell." Dr. Seitz also detected an increasing perceptiveness among viewers, that people are becoming more exacting as they have come under the influence of present-day technical developments. He summarized his views by saying—

... people of today want to be able to grasp quickly what they are looking at. The actual object is no longer sufficient in itself, they also want to learn the facts connected with it—*for it to become alive.* In this respect, members of a museum staff must therefore, if they are to avoid "museum stuffiness," keep in step with the times in their respective branches.

What Dr. Seitz suggests has great professional merit and is an objective which can well be commended to military curators. Even though his colleagues may keenly desire to keep pace with modern museological developments, a great number are still plagued with the frustrations induced by aged and inadequate buildings and other conditions under which they must work. They must thus fall back on their own ingenuity and seek for any devices that will assist them in developing an institution that achieves a measure of high quality.

In many cases a sincere effort has been made to capitalize on certain useful features of an otherwise inadequate building to enhance the attractiveness of an exhibit. The Tower of London provides a notable example. This ancient edifice, whose foundations were laid in the time of William the Conqueror, is used to full advantage in displaying 15th- and 16th-century armor. The armor suits and contemporaneous weapons blend well with the massive architecture of the building. In fact, the building itself is as much a museum curiosity as the armor housed within it. It is not unusual to find the viewer examining the construction of the walls and the beamed ceilings with as much interest as he looks at the various displays within the halls. The Tøjhusmuseet in Copenhagen and the Army Museum in Stockholm are housed in former arsenals. The ground floor in each still resembles an arsenal with long rows of cannon and guns. This form of exhibition is quite effective. Although some of the rooms in the Castel San Angelo in Rome are most inadequate for the display of armor, one might expect to find in such a medieval fortress material of warfare belonging to the same period. Such is the case, and the over-all effect is fairly pleasing.

A number of European military museums introduce the visitor to halls which are crowded with great quantities of artifacts. The greater the number, the less attractive are the displays. Where crowding occurs, the viewer can easily be bewildered by a great quantity of objects all seeking to command his attention. Specific and often excellent specimens are lost amid large numbers of items. The average viewer might well react unfavorably to such displays, feeling that there are just too many objects to look at. If this is the case, it is doubtful whether he will examine many very closely. A tour through such a museum can prove tiresome and frustrating. Yet in each museum there are many excellent specimens which often are displayed well. But this does not offset the possibility that a viewer may get a general adverse impression of the exhibit techniques employed within these museums.

In most museums the objects are well labeled, although in some instances labels are now yellow with age. Dr. Seitz has found that "the general public highly appreciates some form of explanatory text about exhibits, something brief and attractive." Nevertheless, many labels in a number of museums contain an abundance of information which is not needed by any other than the serious student.

At the opposite end of the exhibit spectrum, one finds the spacious halls of the National Maritime Museum in Greenwich and the Army Museum in Vienna. Both museums feature significant and striking artifacts rather than great, extensive collections exhibited in endless monotony. In assessing the high quality of these two museums, it is necessary again to be reminded that both have considerable exhibition space, and both are reasonably blessed with strong financial backing.

The story of British sea power is told in the National Maritime Museum by displaying three categories of specimens within chronological halls. On the walls of each exhibit unit hang paintings of major naval engagements as well as portraits of key naval commanders. Secondly, there are models of the ships which were in the British fleet at the time. Often these are the builders' models and they must be rated as outstanding works of art. The third category of artifacts includes some of the weapons, panoply, and relics of the contemporary period. Only examples of high merit are shown, and there is absolutely no crowding of the exhibits. A great deal of wall space is still open and the floor is not cluttered by a large number of cases. The museum halls are always bright with a combination of natural and artificial light. The exhibit rooms are also models of cleanliness and neatness.

A similar display plan is used in the Army Museum in Vienna. Here the individual chronological halls combine related pictures, maps, standards, weapons, uniforms, and paintings. Again, only significant, attractive specimens are shown. The individual cases are done in the best of taste, even to a point where they are designed in the fashion of the contemporary period. Thus, ornate, carved cases are found in the halls housing objects of the 15th and 16th centuries, whereas the cases in the halls treating more recent periods are in the style of today. For this reason the military museum in Vienna reaches a high point of beauty. There is absolutely no crowding, the halls are spacious, and the specimens themselves are in the highest state of preservation.

The Tøjhusmuseet in Copenhagen, the Army Museum in Stockholm, the Navy museums in Stockholm and Paris, the Leger en Wapenmuseum in Leiden, and the Royal United Service Museum in London comprise another group whose exhibits feature a considerable number of specimens, but there is little or no overcrowding. Individual exhibits are well appointed, and the viewer is given an opportunity to examine an extensive collection without being surrounded by a bewildering array of artifacts. The Tøjhusmuseet has one feature that is unique. In the hall displaying small arms and edge weapons the entire center of the floor is devoted to housing the reference collections. Significant specimens in the collection are placed in exhibit cases, but the complete reference collection to which the exhibits relate is immediately behind the cases in open racks. Thus, an interested viewer may go from the exhibit to the reference collection for further study by walking only a few steps. This arrangement bears considerable resemblance to the open-shelf reference sections of a research library.

Almost without exception those museums which feature spacious exhibition halls and display significant artifacts without overcrowding also have the best individual displays. It is obvious in these instances that the museum staffs are keenly interested in providing the viewer with attractive exhibits and devote much effort to this. The attempt at over-all excellence is carried over to individual displays. The cases themselves are not crowded with specimens and the viewer has the opportunity to examine excellent pieces that are free from the competition of quantity. Swords, pistols, and other objects are so placed that all their features can be easily examined. They also give evidence that the museum has applied the best in conservation techniques. Displays generally make use of manikins which are not grotesque in appearance, and the flags and colors are usually in a very excellent state of repair. Museums which rate very high in the excellence of individual exhibits include the Heeresgeschichtliches Museum in Vienna, the National Maritime Museum in Greenwich, the Army and Navy Museums in Stockholm, the Tøjhusmuseet in Copenhagen, the Musée de la Marine and Musée de l'Armée in Paris, and the Leger en Wapenmuseum in Leiden. These have the most to show the United States in the way of exhibition techniques. Most other museums also have individual displays that merit favorable comment.

By contrast, those museums that tend to be overcrowded carry this condition into individual exhibits. When display cases are crowded, some distinctive features of many worthwhile objects are actually obscured from view. For example, one museum places a border of swords around a number of its pictures, making it virtually impossible for a viewer to examine closely the details of many fine edge weapons. In still other museums, flags and colors are in a poor state of repair. Within a few years many of these valuable historic objects will probably have crumbled away. When discussing the condition of his flags and colors, one museum official observed that this condition is deplorable, but it is now too late to do anything about it.

Several European military museums excel in particular services and have achieved a high degree of technical competence in certain specialized fields. For example, the Tøjhusmuseet in Copenhagen and several others have developed excellent techniques for preserving armor, artillery, small arms, and edge weapons. The Army Museum in Stockholm and the Army Museum in Vienna are leaders in preserving colors and flags. The Army Museum in Vienna is also perhaps the leading military museum in the world in restoring paintings and prints. The Schweizer Landesmuseum in Zurich has one of the most modern and best equipped workshops of any European museum.

Museum directors and curators in many military museums in Europe indicate they are quite familiar with the technical advances occurring in exhibition techniques for military objects. A number state they have a genuine desire to modernize their own displays, but they plead inadequate buildings and a variety of other limitations. It is not certain that a new building and removal of other difficulties would eliminate all the deficiencies that now exist in these exhibits. However, if these changes could occur, the military curator in Europe would at least have the opportunity to experiment with modern exhibition methods.

PROGRAMS AND SERVICES

In common with other museums, the military museum serves two kinds of clientele. The general public visits the museum to be entertained, to satisfy its curiosity and interest about military objects and the relics of military heroes, and to gratify a desire to learn something of the nation's military history as long as this can be accomplished without too much mental strain. The depth of interest shown obviously depends upon the individual. The serious students of weapons and military history also come, but in much smaller numbers. Students view the military museum as a research institution—a source of valuable information which can assist them in the pursuit of their own independent inquiry. They come to the museum to look at the exhibits, but they remain to examine the reference collections at length or to pore over the books, documents, and prints in the museum library and archives. Accidentally or by design, each military museum patterns its activities to meet the needs of both groups. The degree to which each is served is dependent upon the factors of philosophy, staff competence, and adequacy of facilities.

If the objectives of the museum are quite limited and its staff rather small, neither the interest of the public nor of the specialist may be met too well, although it is very possible that the public will have the edge in services. Such a museum's range of activities is necessarily narrow; it will be able to extend only the minimum in services. A museum with a vital program based upon ambitious objectives and staffed by a number of top-flight professionals can give able assistance to both the public and the specialist. In these circumstances the specialist is likely to benefit the most, provided the museum has strong views regarding the importance of its educative functions.

Certain basic services to the public are normally provided by all museums. At the entrance there is usually a sales counter. The visitor is urged to buy a catalogue or leaflet which explains the general content of the museum and suggests a pattern for his tour. The brochure can usually be obtained for a nominal fee.

In general, guided tours are not provided except by pre-arrangement. Even then the tours are conducted for special groups such as students, clubs, or servicemen and women, and a minimum number must be in the party. The tours normally are led by members of the professional staff or, in cases of special visitors, by the director himself. Members of the armed services receive special consideration at all military museums. They are often brought in groups for special lectures and tours. Because of the large number of such groups visiting the Tøjhusmuseet in Copenhagen, the members of the professional staff receive extra compensation for their lectures.

Each military museum is willing to answer questions addressed to it by the general public or the specialist. This is regarded as a normal activity and indeed cannot be avoided if the museum is to maintain good public relations. If a large number of inquiries is received, a great portion of the staff's time is required to answer them. Complicated questions involve a great deal of research and cause the individual curator to complain they prevent him from concentrating on his primary function. Some museums receive a great number of questions while others are spared all but a trickle. The Imperial War Museum counts approximately 11,000 individual communications annually, but many of these are little more than requests for copies of prints. The National Maritime Museum and the Tøjhusmuseet handle 3,000 to 5,000 annually. The Army Museum in Brussels answers about 2,000, whereas the Naval Museum in Madrid receives about 1,000 questions. The Armémuseum in Stockholm records 40-50 major inquiries a year, and the Heeresgeschichtliches Museum in Vienna notes only an average of 250 "scientific questions."

Apart from general services offered to the public, a number of museums have certain special features in their program for visitors. For example, the Army Museum in Stockholm has an occasional Sunday concert in its main exhibit hall as well as a Sunday lecture. The Musée de la Marine in Paris has a monthly lecture sponsored by the "Friends of the Museum." Often a film will be shown during the lecture period. Other museums have an occasional movie also, but these normally are not shown according to a regular schedule. With rare exceptions, specimens are not lent to the public. Most museums will lend books from their library to other government offices and to the military establishments, but the public must read these books in the library itself. Two of the major museums have a restaurant and a visitors' lounge. The National Maritime Museum in Greenwich has a restaurant which advertises its services and encourages large parties to come and enjoy its facilities. The Army Museum in Vienna also has a refreshment room which is capable of accommodating a fairly large number of people.

The assistance European military museums give the specialist or serious student demonstrates their potential as educational institutions. Although small professional staffs performing many other duties considerably curtail activities in this field, enough is done to make the museum's real educational possibilities readily discernible.

The library and reference collections provide the chief sources of assistance for the student. Museums encourage specialists and other interested parties to use these facilities. Often the museum does not employ a full-time librarian, but individual curators and members of the technical staff are usually willing to give their help. This is also true for those who desire to use the reference collections. With the possible exception of the Army Museum in Vienna, all the military museums in Europe aid the specialist in about the same manner. These same services exist in Vienna, but the museum's facilities also include an apartment in which a visiting researcher can live while working at the museum. Other museums encourage visiting students to spend extended periods in study, but none of them are able to furnish housing.

National European military museums usually maintain informal professional relationships with similar local institutions. The director of the Army Museum in Stockholm actually inspects other museums throughout Sweden. Normally this is not done in other countries, and museums deal with one another largely through correspondence or infrequent visits for consultation. Most of the museums lend collections to each other and exchange services and professional advice. They also lend portions of exhibits and individual specimens to military barracks or to other organizations for special occasions where a display of military artifacts is requested. Members of the museum staff are usually willing to consult with the amateur collector who brings objects who have lost their jobs in other public agencies. The museum does not have to budget for their compensation.

Several museums are not placed within the jurisdiction of the nation's defense establishment. For example, the Imperial War Museum and the National Maritime Museum are under the British Treasury, whereas the Armouries of the Tower of London are supervised by the Ministry of Public Works, the agency that looks after national monuments and the maintenance of public buildings. The Navy Museum in Stockholm is attached to the Merchant or Commerce Department, whereas the Army Museum in Stockholm is part of the Ministry of Defense. In this case the Army Museum is the exception to the rule, for it is the only military museum in Sweden under Defense.

In describing the degree of control the government exercises over their operations, most museum directors claim they have little more than budget and accounting matters to clear. Beyond this they feel the museums are pretty much on their own. The matter of budget can be a difficult one, however. As an agency within a department, the museum is in competition with other divisions for the annual outlay and is subject to the vagaries of any economy drives which may occur. However, once the pattern of museum operations is set, the preparation of the budget can become quite routine, except at the point where the museum director anticipates a major expenditure which cannot be absorbed by the usual appropriation or if he desires to expand his museum's activities. In such instances, the attitude of the appropriate minister towards the museum may prove all important in determining what plans and programs can be developed. Accounting for funds expended is the other side of fiscal control and often forms the major portion of the museum director's annual report.

Since governmental guidance is virtually nonexistent except in fiscal matters, European military museums exercise a great deal of autonomy in matters of policy formulation. This function is often shared between the museum board and the director, but there are a number of museums wherein the director has complete control of policy. The prevailing pattern is for the museum to have a board which gives general supervision and guidance and provides the government its liaison point for museum affairs. The board also serves a useful purpose through its ability to exert certain influence upon government officials in behalf of the museum or by acting as a buffer between the government and museum officials. Several military museums do function without a board. In these instances the director cites the advantages he enjoys in having his control of museum operations free from certain restraints.

The controls which individual museum boards exercise differ somewhat. However, they normally reserve only major policy decisions for themselves and accord the museum director a great amount of administrative discretion. Some boards are quite active, meeting once a month, with a possible break occurring during the summer. The boards of the Imperial War Museum, Musée de la Marine in Paris, and the Museo Naval in Madrid follow such a meeting schedule. By convening so frequently these boards are able to keep a fairly close check upon museum affairs and retain a tighter control over policy matters than do those that meet less frequently. The boards of another group of museums meet quarterly and thus exercise a fair amount of supervision, but they are not able to keep in as close touch with routine museum activities as do the others. A quarterly meeting schedule is maintained by the boards of the Musée de l'Armée in Paris, the Sjohistoriska Museum in Stockholm, and the National Maritime Museum in Greenwich. The least number of meetings are held by the boards of the Haermuseet in Oslo and the Armémuseum in Stockholm, which meet twice a year, and the boards of the Leger en Wapenmuseum and the Musée Royal de l'Armée in Brussels, which convene only once annually.

The controls these boards usually exercise include review of the budget, authorization of major purchases, loan of collections, approval of special exhibitions, authorizing lesser building repairs, approval of any organizational changes, and giving suggestions to the museum director. It is only natural that many of the matters requiring board action are presented by the museum director. Most directors express complete satisfaction with the manner in which their boards function and claim they have the full confidence of the members as well as their willing assistance when required.

The museum boards vary in size from 5 to 20 members, the average being from 10 to 12. The larger boards tend to include a number of private citizens who have an interest in museum exhibits of the Royal Navy and those of the Merchant Navy, which are the two principal type displays in the museum. Apart from its Department of Uniforms, Colors, Decorations, and Paintings, the Leger en Wapenmuseum in Leiden divides its work between a section which handles all arms older than 1840 and one which cares for all weapons dating after 1840. The Army Museum in Madrid has no professional military curators on its staff, and so it claims to have no departmental structure. Instead, a lieutenant colonel or major is responsible for each floor, with lesser ranking officers and noncommissioned officers supervising a room or hall.

Several European military museums have an informal unitary organization, and all the principal staff report to the director for assignment of tasks within their field of specialization. Actually, there is a full division of labor in each institution, even though there is no formal recognition of functional departments. Such an arrangement is not too difficult for the director to manage, providing he limits his span of control to his professional staff and key technicians.

The military museums of Europe do not have large numbers of employees. Some directors complain they are undermanned in all departments. Where there is a lack of qualified personnel, a museum cannot undertake a very ambitious program and must limit, to an extent, the services it might perform. Because there is so much to do and often so few hands to accomplish the task, museum staffs work hard. No serious complaints of overwork are heard, although frequently curators comment that they would like to have the opportunity to divert their energies from certain routines which are now unavoidable. Their morale in general seems quite high, and their incentives for professional accomplishment quite unblunted by some disheartening working conditions.

The largest staff is found at the National Maritime Museum in Greenwich, where 75 persons are employed. The Army Museum in Vienna and the Imperial War Museum in London are close behind with 70 and 68, respectively. A second group of museums averages 40 to 50 employees. These include the Army and Navy museums in Paris, the Army Museum in Brussels, the Navy Museum in Stockholm if one tallies the 20 to 25 employees furnished by the government, and the Tøjhusmuseet in Copenhagen. The Navy and Army museums in Madrid follow closely behind with 35. The smallest museum staffs are those of the Leger en Wapenmuseum in Leiden with 18, the Army Museum in Stockholm with 10, and the Army Museum in Oslo with 9. This small number in the Oslo museum can be attributed in part to the fact it was not yet open to the public.

Regardless of the total number of employees at these museums, they all have about the same size professional staff. This varies from three to six curators, including the director, with the Army Museum in Vienna having the largest professional group. The greatest difference in the total number of museum employees is found in the number of specialists, assistants, and technicians. The larger or more pecuniarily endowed museums add greatly to their personnel strength in these categories. For example, the National Maritime Museum, the Imperial War Museum, and the Army Museum in Vienna have 30 or more such employees, whereas several museums have less than 10, and another group has between 10 and 20. The larger museums also have a sizable number of guards or warders. Often these men double as cleaners or perform other menial tasks. Their cost to the museum is usually quite small because many are retired servicemen or pensioners.

Europeans tend to give much recognition and show considerable deference to persons of professional attainment. This is certainly done in European military museums, for great stress is laid upon the curator's academic preparation. He cannot hope to attain professional status unless he has a university degree. Indeed, many of these individuals possess the Ph.D. degree and are lecturers or part-time faculty members at a local university. The deferential treatment they are accorded gives them an added incentive to strive for a high degree of accomplishment.

Recognition is also given to the specialist and technician as a person possessing much needed skills. It is easy to sense that museum directors encourage development toward professionalization within the specialties they require. A curator at the Heeresgeschichtliches Museum in Vienna stressed that the key to his museum's personnel policy is the development of a team concept among the staff members at all levels. Each employee is encouraged to think that his skill is highly important and that a failure to fulfill his potential will lessen the over-all excellence of the museum. The soundness of such a policy is reflected in the in adding valuable items to an already extensive inventory and in improving the quality of the museum's furnishings.

The amount of money a military museum has for its activities is certainly an important factor in ultimately determining its scope and quality. It is obvious, upon careful examination, that the museums with strong financial support are those which are able to present exhibits of the highest quality, employ a very competent staff, maintain a strong program of education, and provide the public with many useful services and considerable entertainment. These museums likewise are able to attain the highest level of professional recognition.

Imperial War Museum, London. The hall displaying models of ships used by the British Royal Navy.

Imperial War Museum, London. Once a hospital corridor, this hallway now displays military objects.

National Maritime Museum, Greenwich. One of the several halls featuring ship models, paintings of the ship themselves, and portraits of their celebrated commanders.

National Maritime Museum, Greenwich. A typical unit exhibit comprised of the painting of a famous ship, and cases containing uniforms and a few objects used aboard.

The Royal United Service Museum, London. A general view of the exhibit hall.

The Royal United Service Museum, London. One of the unit exhibits featuring models of vessels used by the Royal Navy.

The Tower Armouries, London. One of the general halls of arms and armor.

The Tower Armouries, London. The Council Room housing the personal armor of Henry VIII.

Part II: Significant Military Museums and Weapons Collections

IMPERIAL WAR MUSEUM
London, England

The idea for a national war museum developed in England at the end of World War I and took the form of a suitable memorial to the effort and sacrifice of the British people during this great conflict. Apart from its aspects as a memorial, the museum was designed to serve as a repository for war records and become a study center for the broad subject of warfare. The scope of this institution was expanded later to include the memorabilia and other materials from the second World War. The comprehensive coverage of British military operations which is undertaken within the museum is succinctly stated in the opening paragraph of its published brochure:

The Imperial War Museum illustrates and records all aspects of the world wars of 1914-1918 and 1939-1945 and the other operations in which Forces of the British Commonwealth have been engaged since August 1914, and makes accessible to the public and to students data of all kinds relating to them. . . .

With such a statement of purpose, the task essayed is indeed a formidable one. In common with other military museums, this institution possesses custodial, educational, and commemorative functions, but in contrast to many others its coverage for each function is intensive, and a sincere effort is made to attract the interest of all segments of the British population. Because of the broad terms of its legislative authorization, the museum has been able to insure the availability of great quantities of material for exhibition and study. Its broad public appeal has been achieved by exhibiting objects of the Royal and Merchant Navies; the Army and Air Services; the forces of the Dominions and Colonies, the Allies, and the Enemy; the Civil Defence Services; and items related to all aspects of the war effort on the home front.

In the comprehensiveness of its collections the Imperial War Museum, within certain limitations, fulfills the definition of a national military museum. Its exhibits feature all the national military services. While some attention is given to mementos of such war heroes as Lord Kitchener, Sir Douglas Haig, and Sir John French, major emphasis is placed upon objects used by the military establishments in prosecuting the war. Thus, the anonymous veteran can easily identify the weapons and other items he used or saw in mass quantities.

The Imperial War Museum's exhibits date only from 1914. The earlier periods of British military history are covered in the National Maritime Museum at Greenwich, the Royal United Service Institution, the armouries in the Tower of London, and the many regimental museums and specialized collections scattered about the country. The Imperial War Museum likewise gives no coverage to peacetime contributions of the armed forces; nor does any other military museum of prominence in Great Britain.

Like the majority of such institutions in Europe, the Imperial War Museum is not housed in a building originally erected for its use. After tenancy in two prior locations, the museum was brought in 1936 to its present quarters on Lambeth Road in South London. The structure had been originally built as a hospital and so used for over a century before it received the collections of the Imperial War Museum. Therefore, it contains several narrow corridors and small cubicles originally designed to house hospital patients. These areas have been converted to exhibit rooms for the display of military artifacts. The building is susceptible to some renovation, but this would require considerable expense. Thus, the museum staff has had to wrestle with severe limitations and difficult working conditions imposed by a building that has not been modernized or remodeled to facilitate its use as a museum.

The Imperial War Museum has a great quantity of material to display or store in a relatively small building. The inclination is to place an abundance of artifacts on exhibition with the result that the viewer is besieged with a great mass of material in every hall he enters. The museum catalogue includes 600 uniforms, 500-600 medals, badges, and other insignia, 500 small arms, 150-200 swords and other edge weapons, 20 heavy guns, and an armored car, a tank, a one-man submarine, seven military aircraft, a V-1 and V-2 rocket, thousands of ship models, and hundreds of aircraft models.

The vast bulk of this material is on display together with a rather extensive collection of paintings and prints in a space of approximately 40,000 square feet. The museum also contains about 30,000 square feet devoted to ready reference materials, study area, and workshops. Most of this space is allocated for the photograph library which contains some 3,000,000 items and a reference library of more than 60,000 books and manuscripts dealing with the subjects of naval, military, and air history. The library also contains many volumes dealing with the impact of recent wars on the social, political, and economic life of nations. Both libraries are readily available to the public and are probably the most extensive collection of reference materials connected with a military museum in Europe. The museum also sets aside some 25,000 square feet for dead storage and the housing of items that are neither under study nor deemed to be in proper condition for display.

The building design of the Imperial War Museum has made the preparation of attractive exhibits a difficult undertaking. The staff also has found it necessary to place a large number of artifacts in a very small space because the museum possesses a minimum of storage area. Individual objects of great merit are sometimes difficult to locate within the great mass of material available for examination. There is so much of interest to see in each hall that a complete tour is a time-consuming project if an attempt is made to look at every item or to read the many labels in detail. As a result, many viewers might be tempted to give each exhibit a very cursory view and dawdle over only those things which quickly attract attention. Each room contains many cases, and quite often each case is fully packed with excellent artifacts. Throughout the museum cases are placed close together. The corridors often have cases in the center, so that one quickly gains the impression an attempt has been made to utilize available exhibition space to the maximum.

Specimens belonging to each service are grouped together, and there are a number of dioramas to illustrate certain tactics of warfare. Individual exhibits recount major incidents in the two World Wars, but the displays of the museum have not been designed to recount the full sweep of military history for the periods under consideration.

Since the museum attempts to display the bulk of its many collections in a fairly small building, the basic reason for the heavy concentration of objects in its exhibits is somewhat apparent. The staff has also met the problem of insufficient storage space by placing as many suitable objects as possible on display and, in this way, has contributed to the inevitable overcrowding. Museum personnel readily point to certain changes they would like to make. However, the staff is fairly small and individual members have many duties to perform. As a result, they are forced to limit greatly the amount of time they can work on the exhibits. Hence, the displays remain fairly unchanged, except for the addition of new material, an occasional special military exhibition, and an average of three special exhibits a year by the Art Department.

Although the Imperial War Museum has facilities for extensive research, it is not a major research institution. The tasks of the staff are sufficiently time consuming to preclude an ambitious independent research program. Accordingly, no special funds are set aside for it. The director and assistant director are specialists in particular fields and are able to author an occasional publication. For example, the assistant director in recent years has written a superb volume on British uniforms. The bulk of the research performed by the staff is of a routine nature and is accomplished in connection with the museum specimens under study and in answering questions submitted by the public.

In common with every other military museum, the Imperial War Museum is the recipient of many questions from the public. These range from a simple request for the identification of a military object to comprehensive questions about war heroes, battles, and military doctrine. Preparing answers for the sizable number of these inquiries consumes a great deal of the staff's time, but there is little way to avoid this service if the museum is to maintain adequate public relations. The director and assistant director together respond to about 500 such inquiries a year, while the reference library answers 2,500 and the photographic library handles slightly over 11,000. The reference library staff also gives every possible assistance to students of military history and other interested parties who come to the museum for study. The library lends books and manuscripts to such government agencies as the War Office and the Home Department, but it does not extend this privilege to the general public. The museum staff will conduct guided tours for small groups, but only by appointment. The museum does not have the facilities to conduct lectures, symposiums, or classes on military and current affairs topics, nor is it able to show motion pictures on the premises.

The Imperial War Museum was established by Act of Parliament in July 1920 and has always been a government agency. It is under the jurisdiction of the Treasury, and the Financial Secretary to the Treasury is the official who has to deal with all questions relating to it. The assistant director of the museum sees little value in this arrangement, except for fiscal matters. The general management of the museum is in the hands of a board of trustees which has approximately 20 members. Representation upon the board reflects the relation of the museum to the armed forces and other agencies of government as well as the interest of the entire British Commonwealth in its activities. One representative comes from each of the armed services, the Treasury, the Home Office, and the Colonial Office, and one member is appointed from each of the Dominions within the British Commonwealth of Nations. A woman also serves on the board to look after the women's interest in the museum. None of the appointees are required to possess any special qualifications or experience in museum affairs prior to their selection. However, they are expected to look after the interest maintained in the museum by their particular agency or appointing authority. The board deals with matters of major policy, e.g., any large purchases, lending portions of a collection, consideration of any major changes within the museum recommended by the staff, and the annual budget. The Treasury representative has pre-eminence on budgetary matters and gives comprehensive guidance on museum finances. The board gives advice or offers suggestions to the director at his request. It is a fairly active body and, with the exception of twice in the summer, holds a monthly meeting.

The chief executive officer of the museum has the title of director general. He is empowered to exercise full control over all routine administrative operations and must consult the board of trustees in those matters within its purview or where decisions are demanded which appear to be outside the scope of his powers. The director of the Imperial War Museum is more fortunate than a number of his European colleagues, for he is able to share part of his administrative load with an assistant director. The director also has no other professional duties outside the museum.

The work of the Imperial War Museum is divided into five major categories, each of which is handled by a department or division. These divisions are exhibits, the reference library, the photographic library, the film department, and the art department, which looks after all pictures, coins, stamps, and maps. The workshops and all craftsmen employed within the museum are personally supervised by the director general or the assistant director.

At present there are approximately 70 persons on the museum staff. Of these only three are rated curators. These are the director general, the assistant director, and the head of the film department. The extensive reference and photographic libraries require the full-time services of four librarians. These librarians not only work with the materials, but also assist researchers who constantly use the resources of the Imperial War Museum. Most of the technicians in the museum are either photographers or film preservers; there are five in each classification. There are only three craftsmen who work on the exhibits and their chief duty is to condition materials for display. The remainder of the museum employees are the 30 uniformed warders. Their chief duty is to guard the exhibits, but they also perform simple unskilled maintenance tasks.

Several agencies of the British Government share the total cost of operating and maintaining the Imperial War Museum. Parliament appropriates about half the money required directly to the museum for application to the payment of salaries and the cost of other routine operations. The remainder of the total cost is absorbed within the budgets of those departments which are obligated by statute to provide the museum, as a government agency, with certain services. For example, all repairs and major maintenance expenditures for the Imperial War Museum building are budgeted by the Ministry of Works which performs a similar function for all government-owned structures. Other offices provide the museum with its stationery, pay its utility bills, and contribute allotments for employees' pensions. The museum does not charge for admission, but it does realize a profit from the sale of publications, photographs, film rights, and photo copyrights. This income is not treated as added revenue for museum use but reverts to the Treasury and is credited toward the cost of operating the museum.

The Imperial War Museum attempts to maintain friendly professional relationships with the other British military museums. Normally it does not lend its collections to these museums, for each has its own high degree of specialization which in general does not overlap that of the Imperial War Museum. It is the museum's policy to give specimens which it cannot use to other museums which in turn reciprocate. The curators and other specialists of British military museums regularly consult with one another on matters of mutual professional interest.

In conclusion, the Imperial War Museum meets the interests of both the specialist and the general public. The collections are sufficiently specialized to attract those who have interests in specific types of weapons and other military objects. On the other hand, a British soldier or sailor can enter the museum and reasonably expect to find some object or perhaps many items which may have belonged to his particular unit. The number of visitors to the museum is probably curtailed to some extent because of its location. It is easily reached by public transportation but is sufficiently removed from the heart of the London tourist area to discourage many who are not genuinely interested in looking at military collections. In spite of certain liabilities with respect to location, the Imperial War Museum attracted 245,000 visitors in 1957.

NATIONAL MARITIME MUSEUM
Greenwich, England

The National Maritime Museum at Greenwich is one of the most spacious and best appointed military museums in Europe. Created by an act of Parliament in 1934, this museum was founded "to promote and maintain due interest in British Maritime History—the work of the Royal Navy, Merchant Navy, Fishing Fleet, Explorers and Yachtsmen; and to collect and preserve, to exhibit or make available to all who are interested any objects which explain the story or assist the student." In Europe there are only two other naval museums that appear to attain the grandeur and scope comparable to the National Maritime Museum. These are located in Stockholm and Paris.

This museum is the one major English military museum that is not in the heart of London. The setting, however, is a natural one, for it is located adjacent to the Royal Naval College near the bank of the Thames River. This site is not only the repository of much naval lore but has been the scene of many interesting episodes in British history. Here, at one time, was the summer home of British monarchs. This was the birthplace of Queen Elizabeth I, and it was at this location that Sir Francis Drake reported the results of his voyages to the Queen. One of the museum buildings is the famous Queen's House built by James II and Charles I. Thus, its historic setting assists in making the National Maritime Museum a prime tourist attraction.

Three separate buildings, none of them new, house the museum collections. The central building, known as the Queen's House, was built in the mid-17th century. The adjacent wings of the museum are located at some distance from the Queen's House and are connected to it by colonnades. These were all built at a considerably later date. The two side buildings are long and narrow but are ideal for museum halls. They offer no insurmountable obstacles to the display of any military specimens the National Maritime Museum possesses. The general excellent repair of the buildings indicates they were renovated at the time the museum was founded. They have been kept in excellent condition since that time, and so there is a general impression that the museum is a new one.

Most of the museum's specimens are not physically large. The inventory includes 550 uniform garments, 600 badges, 3,400 insignia and medals, 82 small arms, 160 naval swords, 50 other swords, 24 cutlasses and 12 pikes, 33 guns from ships and boats, 6 full-size boats, a large number of anchors, 30 engines, and approximately 1,000 ship and boat models. Also carried on inventory are 1,150 navigation instruments, 5,000 charts, 3,300 oil paintings, 25,000 prints and drawings, 100,000 photographs, 45 ship draughts, and about 1,000 additional objects related to key naval heroes. Most of these consist of furniture, china, silver, and miscellaneous relics together with uniforms, swords, orders, and medals belonging to Nelson, Duncan, Franklin, and other lesser naval leaders. The museum also contains a library of 38,000 books and 10,000 volumes of manuscripts.

The National Maritime Museum is indeed fortunate in having ample space to display its extensive collections. The Queen's House, which must also be considered an exhibit area, does not contain military specimens. There are a few pieces of furniture located in the building itself, and the walls are hung with a number of paintings; there are insufficient paintings and other objects within the building to detract from the beauty of the rooms and the general attractiveness of the house itself. The Queen's House is considered one of the finest achievements of the famous architect Inigo Jones, the same man who designed the Banqueting House at Whitehall. Thus, the museum artifacts are displayed in the remaining two buildings.

The interior exhibit space totals 96,565 square feet. There is a small outside area of approximately 900 square feet. The ready reference materials, which include all reference collections and objects under study, are housed in a space of 9,400 square feet. The museum also carries about 10,000 square feet in the category of dead storage, although it is often difficult to draw a line between ready reference and dead storage. The assistant director of the museum has stated that all material not on exhibition is to a varying degree available for study. In his opinion, a museum should never have any dead storage and all materials not needed should be disposed of. The library occupies a space of 5,500 square feet. The books and manuscripts are kept in the finest state of repair because all the library bookshelves are completely enclosed. There are study rooms adjacent to the library and these contain 3,530 square feet. The museum also has a lecture room of 1,876 square feet. About 3,000 square feet are allocated for office space, whereas the scientific laboratories and maintenance shops are contained in 2,150 square feet of space. The museum also has several internal service rooms totaling about 2,500 square feet.

It is estimated that the corridors and stairways amount to an additional 22,000 square feet of space.

With the exception of several special galleries, the museum's exhibits are arranged so that the visitor is able to progress chronologically from one period of British naval history to another. For example, one hall displays naval objects coincident with the reign of Queen Anne; another hall covers the Seven Years War; yet another is devoted to Cook's voyages. Other halls treat the naval campaigns of the American Revolutionary War, the French Revolution, the Battle of Trafalgar, and one displays objects relating to Lord Nelson. Several halls exhibit specimens from the two great wars of the 20th century. In addition to this chronological presentation, a series of halls cover the evolution of the Merchant Fleet from sail to steam. One large area, which had previously been a gymnasium, is used to display large ship models and the figureheads from important ships. Another room is devoted solely to the display of navigation instruments. Still another contains the museum's large collection of naval prints and special ship draughts. There is also a small room set aside for the display of medals and one for naval seals.

The chronological halls show excellent planning by the staff. A particular period of naval history is illustrated by grouping together pertinent ship models, paintings of naval engagements, and portraits of the naval heroes of the day. In addition, there are cases containing contemporary relics such as ships' logs, china, books, or other objects which may have been used aboard ship. Because Lord Nelson figures so prominently in British naval history, he is the only hero who is memorialized by a complete hall devoted to the exhibit of his personal objects or those related to his exploits.

The museum possesses a magnificent collection of oil paintings by two famous naval artists, both named Van de Velde. Because of its extensive collection of paintings, the National Maritime Museum has had to adopt the display techniques employed in a good art gallery. In exhibiting its paintings, the museum has had to cope with the problem of limited wall space created by the large number of windows in its buildings. It has solved the difficulty by erecting partitions perpendicular to the walls. These partitions reach only partway to the ceilings and are only wide enough to contain one of the large Van de Velde paintings or the portrait of a naval hero. Thus, each individual hall appears to be broken into a large number of alcoves with the space between the partitions used for cases containing specimens related to the period under consideration. All halls have wide aisles, and these contain several display cases which not only provide an attractive setting for the exhibition of objects but also insure effective space utilization within the halls themselves.

With the possible exception of the one large topical hall containing ship figureheads and some of the large models, there is absolutely no crowding or clutter. Since only significant objects are displayed in the museum, the viewer is not wearied by having his attention drawn to a huge quantity of specimens. In one of its brochures, the museum contends that a complete tour of its facilities in one visit is rather strenuous, for it entails over a mile of walking. It advises its viewers that an itinerary should be planned beforehand so that portions of the museum can be omitted to suit the age and interests of the party.

The visitor to the National Maritime Museum is impressed by the high quality of the exhibits and the excellent artifacts. He is able to examine one of the finest collections of marine paintings in existence and a superb collection of models constructed by the builders of the original ships. The theme of patriotism is not overplayed, but the viewer is constantly reminded of the past glories of the British Navy, and he is able to see through graphic presentation the role that British sea power has played in international politics.

The National Maritime Museum has greatly benefited from a large number of splendid donations. For example, the Royal Naval College transferred its Admiralty collection of pictures, relics, and ship models. The Admiralty has transferred to the museum the official draughts from which the ships of the Royal Navy were built between 1700 and 1900. Lord Sandwich presented his collection of naval medals and several members of the Nelson family have contributed a large quantity of the museum's Nelson relics. The greatest single donor to the museum was Sir James Caird. Every department of the museum benefited from his presentations, for he possessed the largest single collection of naval materials in Great Britain. His objects included pictures, models, instruments, books, and other relics. It was largely through the influence of this benefactor that the museum took its present form. He also left a trust fund at the museum's disposal.

In general the services the National Maritime Museum provides the public conform to the pattern performed by other military museums. The staff undertakes some research, but most of it deals with the military specimens in the museum inventory. Individual members of the staff seldom produce studies of their own. However, the director, deputy director, and some of the professional employees write occasional articles. Two volumes produced at the museum are of considerable professional interest. One discusses the various types of uniforms used by the British Navy throughout its history and another book describes and illustrates the Van de Velde paintings. The staff must also conduct a certain amount of research when answering those questions received from the public, which number 3,000 to 5,000 a year. The staff is also willing to identify objects brought to the museum by interested parties and will assist any amateur collector of naval materials. These services are performed on an informal basis.

The museum also performs certain educational functions. During the summer it works with the Ministry of Education and every year conducts a 10-day course on maritime history in conjunction with the Royal Naval College. Members of the staff lecture to the students who are assembled for the course. The museum is also the scene for the annual meeting of the Society for Nautical Research. During these meetings a lecture is given at the museum by an outstanding naval authority. Apart from these two annual affairs, no other lectures, seminars, or special courses are conducted at the museum; the staff is too limited to engage in any more ambitious educational effort. These formal educational activities are augmented by an occasional guided tour. School groups are urged to visit the museum. When they do, they are usually escorted by a member of the professional staff or are assisted by warders stationed in the principal halls. The warders are qualified to answer certain questions and to furnish the viewer with general information about the objects in the halls under their care.

The museum has a restaurant which seats 80 persons in its main room and about 30 on an adjacent terrace. In favorable weather, luncheons and teas are served on the lawn. The museum advertises the services of its restaurant and encourages large parties to come and use these facilities. The restaurant functions only during the summer months.

The National Maritime Museum is a government agency but it is not controlled by the Royal Navy. Instead, it comes under the general administration of the Treasury. In this type of control it is similar to the Imperial War Museum. However, the extent of Treasury supervision is limited to financial matters.

Since its founding, a board of trustees has formulated major policy for this museum. The Act of Parliament which created the museum specified the Treasury as the appointing authority for members of the board. However, by custom, they are named by the Prime Minister. Twelve eminent citizens serve on the board, and appointment is for a period of seven years. An individual may be reappointed at the expiration of his term. The present chairman of the board has been a member since the founding of the museum. Apart from establishing major policy for the museum, the board authorizes all important purchases and any loans which the museum wishes to make of any part of its collections. The trustees interfere to some extent with museum operations, but normally such interference is confined to matters in which the museum might get into difficulties. Individual members offer varying degrees of assistance to the staff and to the director and make suggestions based upon their observation of museum operations. The board meets four times a year. One of the ablest members in the opinion of the staff is the Duke of Edinburgh, the Queen's husband. As a professional naval officer, he displays an interest in all aspects of museum operations.

The substantive workload of the museum is accomplished by four major departments: the Department of Models, Draughts, and Relics; the Department of Pictures; the Department of Navigation; and the Department of Manuscripts. The library is also listed as a separate entity in the museum structure, and there is a service type department which may be conveniently called a Department of Administration. Included in this department are the clerical personnel, warders, and accounting section. As in other museums, the administrative personnel are supervised by the director or deputy director.

The National Maritime Museum in Greenwich has one of the largest staffs of any military museum in Europe. It normally employs 75 persons, all of whom are civilians. However, the deputy director is a retired naval commander and is regarded as a major authority on matters related to the Royal Navy. Three persons on the staff are designated as assistant keeper, first class, which corresponds to the rank of curator. Each one of these is a department head. The librarian is also an assistant keeper, first class, but the head of the Department of Manuscripts has the rank of custodian, which is at a slightly lower salary scale. The deputy director has a research assistant. There are also research assistants in the Department of Models, Draughts, and Relics and the Department of Pictures. With the exception of the Department of Navigation, the major departments also employ one or two craftsmen. Within the Department of Administration there is a section called the Establishment Office. It is headed by an individual who is rated as a higher executive officer. Apart from certain clerical help, the Establishment Office includes a master craftsman and two photographers. The museum has 9 clerk-typists and 32 warders. Included in the clerical staff are three people who work in the finance or accounting office. Completing the museum's employees are 12 janitors or cleaners and 2 messengers. During the summer the museum also employs 6 temporary warders.

Arrangements for financing the operations of the National Maritime Museum are almost identical with those for the Imperial War Museum. Parliament appropriates about 40 percent of the cost directly to the museum for salaries, preparation of exhibits, and some new acquisitions. The Ministry of Works receives slightly more than 50 percent to cover the cost of maintenance, furniture, fuel and light, and other services it performs for the museum.

Largely because of its strong financial base the National Maritime Museum has been able to develop into one of the best military museums in Europe. Its contents are of the highest quality. In part this has been assured because the museum has benefited so greatly from the donations of Caird and other eminent persons. The benefactors of the museum have also been much interested in having their collections housed in the finest surroundings.

The location of the museum in Greenwich does not appear to prevent it from attracting a large number of visitors. Greenwich itself draws many tourists because it houses the Royal Naval College and other historical objects of interest, and because it is the site of the Greenwich Observatory and meridian. Since the observatory has recently been renovated to make it an adjunct of the museum, it is possible that the number of museum visitors will soon exceed the present average of 400,000 per year.

TOWER ARMOURIES
London, England

One of the major tourist attractions of the Tower of London is its rather extensive collection of armor which is located within the oldest of the Tower buildings. The collection has approximately 8,000 items in its inventory. Most of the armor is in the form of complete suits or separate pieces such as breastplates and helmets. There are also a large number of small arms and edge weapons. Most of the armor dates from the 16th and 17th centuries, but the arms collection includes specimens as recent as 1914.

The Tower Armouries cannot be considered a national military museum within the full definition of the term. It is principally a specialized collection of armor housed in an edifice that appears contemporaneous with the arms on display. Since many of the objects on exhibit were used at a time when the Tower was a center of much royal activity, their presence is quite natural within a building whose construction was begun by William the Conqueror. Most assuredly the building was not originally constructed to house a collection of armor. Nevertheless, its large halls have been readily adapted to an attractive display. They have not had to be remodeled to exhibit the armor but have required only the addition of the requisite appurtenances for displays and suitable artificial lighting.

Because the Tower of London is of such historical significance it possesses many of the attributes of a museum. The Tower buildings and, in particular, the one housing the armor collection, are as much of interest to the visitor as the armor itself. It is not unusual to see visitors examining the walls, ceilings, circular stair wells, and gun slots as closely as they do any part of the armor collection. Therefore, the exhibits are in a very natural setting, and it is evident that the keeper of the armouries has tried to capitalize on this asset to make his displays attractive.

The space in all halls is used efficiently. None gives the appearance of being crowded or cluttered. Many of the complete suits of armor are displayed to advantage on manikins. A great deal is also done with a realistic display of horse armor and most of the horses are mounted with armor-clad riders. The armor is in an excellent state of preservation, reflecting superb care by the museum staff. The showcases are relatively modern and in a good state of repair. In those cases which display swords and small arms, the glass is angled in an effort to avoid the glare emanating from fluorescent lighting overhead. All labels are attractively hand printed in a modified Old English script. Their narrative descriptions are relatively short but lucid.

The museum office and workshops are not in the same building with the collections, but are in an adjacent structure which was acquired in the past few years and is now being renovated to accommodate additional exhibits and better facilities for the staff.

The Tower lacks many of the facilities that normally appear in a museum. For example, it has no library apart from the books located within the offices of the keeper of the armouries. There are no internal service rooms, security vaults, receiving areas, exhibit laboratories, or study rooms. The workshops are comparatively small and are equipped only for the conditioning of the armor, construction of cases, and any other work that must be performed in readying the specialized objects of the collection for exhibition. Storage space is at an absolute minimum. One reference collection is kept in storage for loan purposes, and sufficient other pieces of armor, guns, and edge weapons are kept in ready reference for the serious student to examine. Hence, most of the specimens in inventory are displayed on the three main floors of the original Tower building and in the storerooms or vaults beneath the building.

The Tower Armouries can provide little more for the public than a display of the collection. Hence, the museum's primary function is that of custodian, and the staff is always on the lookout for additional pieces with which to fill the gaps in its collection. Any educational function it performs is only in the form of a byproduct of the exhibits themselves. The staff is also willing to identify objects and answer questions asked by the public concerning weapons and armor. There are not sufficient people on the staff to provide guided tours, but the labels for the objects are sufficiently comprehensive to preclude the need for any extensive oral explanation. Signs direct the visitor through the museum in a unidirectional flow of traffic.

Only a small staff is needed for the museum because its activities are limited to displaying the objects in the collection, and the exhibits themselves remain quite static. The supervision of the collection is in the hands of a person who has the title "keeper of the armouries." He also manages the extensive Wallace Collection of armor, which is housed at another location within London. Most of his time is spent at the site of the Wallace Collection so that the actual management of the Tower Armouries is now in the hands of his assistant. The Tower also has one professional armor expert who is rated at the rank of curator. The remainder of the staff includes three or four technicians who work with the materials and a number of uniformed guards who are located in every hall of the museum.

The entire Tower of London is managed by the Ministry of Works, the Armouries comprising a department within the over-all Tower organization. For this reason, the Armouries has no board of directors, but the Ministry of Works may interfere in its operations at any time. Since the Ministry of Works has rather extensive responsibilities covering a variety of Government enterprises, it has adopted the policy of not interfering in the technical activities of the Armouries. It limits its concern to repairs and construction work on the buildings and includes the operating funds for the Armouries in its general budget. The routine expenditures for the Armouries such as salaries, materials, utilities bills, and repairs are all incorporated in the financial statement of the Ministry of Works without reference to any particular department or section within the Tower of London. However, the Armouries are specifically allocated a certain sum each year to spend for the acquisition of new specimens. No charge of admission is made for the Armouries, but there is a charge for entering the Tower of London. The funds collected for admission to the Tower are not applied specifically to maintain the Armouries. The Armouries likewise receives no profit from the sale of publications, souvenirs, or other articles.

Apart from providing the visitor with the opportunity to see one of the finest collections of armor in Europe, the Tower Armouries reaches one of the highest points in entertainment value among military museums. One is perpetually reminded that he is on historic ground and at the site where significant events occurred in centuries past. The suits of armor worn by Henry VIII are perhaps of more interest to the viewer when he realizes that this celebrated British king undoubtedly knew the Tower buildings well and had frequently been there. The visitor's contact with English history continues wherever he goes throughout the Armouries but reaches a high point when he enters the beautiful Norman chapel where Knights of the Bath kept their vigil and the youthful and ill-fated Lady Jane Grey knelt in prayer during her brief reign in 1553. Thus, vivid impressions of a rewarding experience remain with the visitor long after he has left the Tower Armouries.

ROYAL UNITED SERVICE MUSEUM
London, England

The Royal United Service Museum is located in one of the most historic buildings in London, the impressive Banqueting House of old Whitehall Palace, opposite the headquarters of the famed horse guards in Whitehall. The museum is part of the Royal United Service Institution, which was founded by William IV in 1831 and now possesses some of the great treasures of the British Navy, Army, and Air Force. Although the Institution has existed since 1831 the museum was not founded until 1890. In that year Queen Victoria gave the R.U.S.I. the privilege of using the Banqueting House to display its collections of military objects.

The Royal United Service Institution is an organization created solely for the commissioned officers of the Armed Forces of the Empire and Commonwealth. Membership can be acquired only after approval by the board of the Institution. The R.U.S.I.

provides its members with a reading and writing room containing current periodicals, newspapers, and the principal service publications. It also contains a service library, and members of the staff give advice on professional reading and research. The Institution has an auditorium where lectures are given by eminent authorities on subjects of interest to all three services. Field Marshal Montgomery's much publicized criticism of United States World War II strategy and post-war foreign policy was delivered during October 1958 as part of an R.U.S.I. lecture series. These lectures normally are followed by discussions in which officers of every rank are encouraged to take part. The museum is an integral part of the Institution and provides a place where the members may come to study objects of military interest. But the museum also welcomes visitors and indeed depends upon them for a share of the income which supports it.

The chief executive officer of the Royal United Service Institution is its secretary, and his duties include general supervision of the museum. By training he is not a museum specialist, but the present secretary has given considerable thought to what the R.U.S.I. Museum should do, both for the public and for its members. When he first examined the role of the museum in the total enterprise, he recorded many of his thoughts and from them he fashioned the museum's present philosophy. A few of his comments are of interest, not only in understanding the approach taken at the Royal United Service Museum, but also because they embrace certain fundamentals which confront every museum in choosing its objectives.

First of all, the secretary stated that a museum must select and define its field and aim to give the best possible service within it. Then speaking specifically of the Royal United Service Museum he pointed out that the view of the museum's function must not be too austere. He feels the museum is designed to interest and educate the public in the history of the three services by displaying objects with a natural attraction. He believes it is impossible to satisfy the taste of every individual, but when a point of view is held by a large number it should not be brushed aside.

The secretary notes that the Royal United Service Museum is comparatively small. Although the materials it contains are national in scope, the museum really makes no effort to become a large national institution. There are grave limitations of space, and while it would be highly desirable to give comprehensive coverage to the many facets of British military history, it must be recognized that this is virtually impossible. In summarizing his philosophy of the Royal United Service Museum, the secretary has written, "Museums, like human beings, are better for restricting their diet to what can be digested; both readily display the effects of gluttony."

The Royal United Service Museum is favorable located in the center of the tourist area of London. It is midway on Whitehall and within easy walking distance of the Palace of Westminster or Parliament. The Banqueting House is all that remains of the old Whitehall Palace. It is an architecturally splendid building, having been designed by the celebrated Inigo Jones, and takes the form of a double cube. Perhaps the greatest attraction in the building is its ceiling, which consists of magnificent and massive paintings by the Flemish master Rubens who was commissioned for this work in 1629 by Charles I. In designing his display the secretary of the Royal United Service Institution has stated that the Rubens paintings must be the center of attraction. Hence, he refuses to hang flags or locate other objects where they might obstruct the view of the ceiling.

The building has undergone few renovations to convert it to a museum. In its original design the hall was spacious, and so there has been no problem of removing partitions to make adequate exhibit space available. The museum is housed on two floors. The main floor is about two stories high, whereas the lower floor of the museum is really the basement of the building. It has a fairly low ceiling and, in the manner of most basements, is inclined to be somewhat dark. The secretary states that for architectural and sentimental reasons it is essential to keep the character of the Banqueting Hall and to maintain the tradition of the building. He further points out that the Institution was granted the use of the Banqueting Hall by Queen Victoria and that this privilege has been graciously continued by the Sovereign. The full conditions of the tenancy, therefore, he thinks, should always be borne in mind when considering the layout of the museum or any proposed changes.

Wall space on the main floor is considerably limited because of a large number of recessed windows. Hence, it is impossible to conserve floor space by placing the exhibit cases against the walls. Nevertheless, the museum does make some use of the alcoves created by the recessed windows to display paintings or other small objects. As a result, the museum appears slightly crowded, although there is ample space to move around the exhibits and to get a good view of them. The exhibits are grouped to tell a particular story or incident, but there is no effort to describe the broad sweep of British military history. Important specific incidents displayed include the execution of Charles I, a very excellent panorama of the Battle of Trafalgar, and objects related to major campaigns by the Duke of Wellington. All objects are well and attractively labeled so that the viewer has no difficulty in understanding the exhibits.

The secretary of the R.U.S.I. has given much thought to the matter of exhibits. He believes that objects can always be arranged according to a specific idea or theme with each small section telling a short story. However, the emphasis in telling the story should not defeat the design. He further believes it is neither possible nor desirable in a museum of this size to show the complete evolution of a particular weapon or a military epoch in one specific type of exhibit. In any event, such a display would not appeal to the general public upon whom his museum must depend for its income. He states that it is possible to make an exhibit more comprehensive by supplementing military objects with pictures, charts, and maps, but in any display it is preferable to show only the most interesting and beautiful specimens and not exhibit too much. He adds that he thinks most people find it very tiresome to see a large number of swords or firearms at one time.

Apart from the museum specimens themselves, the Royal United Service Museum possesses one of the finest collections of dioramas in the world. These are fairly large and depict major battles, not only those of importance to British history, but also those relating to the military history of the world. These include the Battle of Hastings in 1066, the Storming of Acre in 1191, the Battle of Crecy in 1346, the Battle of Marston Moor in 1644, and the Battle of Blenheim in 1704. These dioramas are well illuminated and provide an example of the highest form of this type of art.

Because the museum is an integral part of the Royal United Service Institution, its organization is comparatively simple. The secretary establishes the broad lines of policy and must approve any major changes which take place within the museum. The daily operations are in charge of an assistant to the secretary who devotes full time to the care of the artifacts and to all other details of museum management. He is assisted by a fairly small group of employees. Inasmuch as the philosophy of the museum requires that its field be limited, the size of the staff employed is commensurate with the museum's restricted outlook.

The Royal United Service Institution is financed largely by public subscription and membership dues. The museum as an integral part of the Institution must thus be considered a private museum rather than a public one. A portion of its operating costs is paid from income derived from admission charges and from profit realized from the sale of publications, postcards, or other materials. Its sole relationship to the British Government stems from its location on Crown property, free occupancy of which has been granted by the Sovereign.

MUSÉE ROYAL DE L'ARMÉE
et
D'HISTOIRE MILITAIRE
Brussels, Belgium

Le Musée Royal de l'Armée et d'Histoire Militaire first originated as part of the 1910 World Exposition in Brussels. At that time a sizable collection of Belgian Army souvenirs was gathered and displayed as one of the exposition exhibits. When the fair had concluded, the decision was made to retain the collection and house it in one of the buildings that still remained intact. This later became known as the Palais du Cinquantenaire and is still the location of the museum.

The name given the museum would indicate that the exhibits are devoted only to army objects. Actually all services are represented in the collections and the museum in Brussels tends to fulfil the role of a national armed services museum. The Army has been the chief armed service in Belgian history, and it is natural that it would predominate the museum exhibits. The Belgian Navy has never been very large and the Air Service was not organized until just before World War I. Belgium capitulated within a few days after invasion by the Germans in both World War I and World War II, and so the Air Service had little chance to prove itself in combat in either instance. By contrast, the Army has been a very active and large force since the Belgian Nation attained its independence in 1831. This is reflected in the museum through its large collection of army materials and extremely limited ones for the Navy and the Air Service. The museum collections are not limited to Belgium alone, and one can find a sizable quantity of material from nations with which Belgium has fought or been allied. For example, there are items belonging to American military heroes of World War II, and a rather extensive collection of United States materials from World War I, as well as objects from several other countries.

The philosophy of the museum is rather easy to discern from its contents and method of display. Every effort appears to have been made to gather a great quantity of military artifacts and to display as many of these as possible. The apparent intent is to make this institution a great repository for worthwhile military objects, and it is quite obvious that its custodial function is given the greatest emphasis. The museum has been fortunate in acquiring a great quantity of worthwhile specimens, and so the viewer is quickly impressed by its size and scope; it tends to be one of the largest military museums in Europe.

In many ways the Palais du Cinquantenaire is an ideal building for a museum. It is located adjacent to the Cinquantenaire Arch, one of the major landmarks of Brussels. The exterior of the building is impressive and is matched by an almost identical building on the opposite side of the arch across the boulevard. The museum is also very spacious and has ample room to display a great mass of material as well as objects of considerable size. Its facilities presently accommodate some 4,000 heavy guns, 4 armored cars, 190 pieces of mobile artillery, 4 tanks, and 18 airplanes. In addition, the museum displays some 500 complete uniforms, over 7,000 pieces of uniforms, 6,000 swords and other edge weapons, some 5,000 insignia and medals, and 500 small arms. The museum's inventory also includes some 2,500 munitions, over 400 colors and flags, about 250 musical instruments, 50,000 prints, 50,000 photos, 12,000 maps, and 1,500 watercolors, drawings, and oil paintings. Added to these are a considerable number of uniforms, decorations, arms, and souvenirs belonging to Belgian kings and famous generals.

A few measurements will give some key to the museum's size. It has 112,000 square feet of exhibit space, 2,700 square feet of ready reference, and 4,300 square feet devoted to dead storage. The library has 4,850 square feet of floor space and contains some 26,400 feet of shelves for books and about 8,250 feet of shelving for archives. The museum has two study rooms each of which is 860 square feet in size. The maintenance shops have 1,600 square feet of floor space, and the other internal service rooms cover an area of 4,000 square feet. The museum also has a fairly large space for temporary storage amounting to 7,500 square feet and an auditorium of 10,000 square feet.

With such a large building to house its collections, it might be anticipated that the museum exhibits would not be crowded. However, upon entering the museum, the viewer is immediately besieged with a tremendous quantity of material and must decide which of two parallel, curved halls he should tackle first. As he proceeds down either of these lengthy halls and finds there are still several others to visit, he soon discovers that the ample contents of the museum cannot be thoroughly examined in a comparatively short period of time.

It is doubtful if there is a more concentrated display of military objects elsewhere in Europe. Each hall contains several rows of cases and each case is packed with specimens which are, for the most part, well labeled. Cases line the walls, and they too are filled with objects. Artillery pieces and large guns are placed between the cases. A number of halls have flags jutting out from the walls close to the ceiling. Other halls have airplanes suspended from the ceiling, and there is hardly a bit of wall space that is not covered by either a portrait, a picture of a battle scene, or a bust of some Belgian military hero. Some of the busts and pictures are completely surrounded by swords or guns which jut out from behind and give the appearance of an extra frame. This provides an attractive rosette type design but is hardly conducive to a thorough examination of the pieces. Such a display does have one virtue; in part, it solves the storage problem for edge weapons and small arms. With such a huge quantity of artifacts to see, the viewer might properly decide to visit the museum a number of times if he hopes to give all its collections a thorough examination. Because so many objects are displayed, very important and lovely pieces are quite often not isolated from the great mass of specimens and might be overlooked in a hasty tour.

The comprehensive coverage of the museum is impressive, and most of the objects on display are in excellent condition. Many items of considerable historical importance are also on display. Perhaps the most striking thing to be seen in the museum is a gigantic panorama of the Battle of I'Yser, measuring 375 feet by 50 feet. The halls are divided topically and chronologically. For example, one section displays military objects in use when Belgium was part of France, and the exhibits in another section of the same hall deal with the period when Belgium was part of Holland. One hall covers the period 1831 to 1914 with individual sections devoted to exhibiting materials belonging to King Leopold I, King Leopold II, King Albert I, and King Leopold III. One very large hall displays technical objects, one presents exhibits of World War II, and another displays trophies of the First World War. Adjacent to the latter is a hall containing objects which belonged to Belgium's allies in the First World War.

The principal exhibits change very little. However, the museum does give special exhibitions from time to time, but the staff is probably too small to do very much in this particular field. The museum also will display private collections, and one hall is set up for temporary expositions of this nature. When new material is received at the museum, it is normally incorporated into existing displays. Yet acquisition of an entire collection might prompt a special exhibition.

A review of the services performed by the museum demonstrates the emphasis given to the custodial function. The staff conducts no special research projects. It is principally engaged in the upkeep of the displays and caring for specimens. Even though the staff does not spend much time in personal research, a very excellent library serves as a resource for students, and members of the staff are available for assistance in their particular field of specialization. The museum likewise gives no special lectures for the public. However, when regiments of the Belgian Army pass through Brussels they often come to the museum to view the exhibits and hear lectures on Belgian military history given by the staff. Guided tours are normally not available at the museum. However, arrangements can be made with the staff to conduct parties of students through the museum. The staff also answers about 1,500 to 2,000 inquiries per year addressed to the museum, and they are willing to identify specimens and provide other information requested by amateur collectors. To do much else for the public would require additions to the present staff.

The Musée Royal de l'Armée is a government agency. It is part of the Ministry of Defense, but the Ministry does little more than to provide for its budget. The museum is governed by a policy-making body called the Commission of Surveillance which has a membership of 10 to 12. The chiefs of the major divisions of the Ministry of Defense, that is, the chiefs of Aviation, Army, and Navy, serve as members of the board, as do representatives from each of the lesser divisions of the military establishment. All members of the Commission are military officers on active duty. None possess any special qualifications in museum matters, and their service on the Board must be considered an additional duty. The Commission of Surveillance meets only once a year. It comes to the museum to make its annual survey and after a trip through the premises with the director makes suggestions, which the director generally tries to follow. The commission also examines new acquisitions, considers what it would like to do in expanding museum facilities, and analyzes the costs of operation. It issues no orders; it only offers suggestions.

The top executive official of the museum carries the official title of chief curator or conservateur en chef. He has virtually complete control of daily operations and, with the exception of his annual consultation with the commission, he is on his own. The museum has no assistant director, but the director does have an administrative assistant. In the half century of its existence, the Musée Royal de l'Armée has had two directors. The present director succeeded to the position held by his father since the founding of the institution.

There are currently 50 members on the museum staff. Of these, only four are classified at the professional level—three curators and one librarian. The technicians employed include two individuals who work solely on exhibits, two carpenters, two painters, one person working on labels, an electrician, an ironworker, and several general duty laborers. There are also 10 guards and 12 members of the janitorial staff. For reasons of economy, the armed services provide most of the maintenance people and the janitors. The staff also includes one stenographer.

The museum's organizational structure is comparatively simple for there are only two major divisions: One handles the substantive functions of the museum and includes the professional and technical members of the staff; the other is concerned with administrative matters and generally embraces the activities of the administrative secretary, the guards, maintenance people, and the janitorial employees. The simplicity of this form of organization, coupled with the limited size of the staff, permits the director to exercise close supervision over the wide range of museum activities. Such administrative conditions also make it very easy for him to impart his strong personal enthusiasm about the museum directly to all members of his staff. Indeed, in his approach to the broad problems of management, the director easily conveys the impression that his earnest desire to enhance the museum's reputation is strongly motivated by great family pride in this institution.

The museum is fully dependent for its financial support on the national defense budget, but specific details about the allocation of funds are difficult to obtain. Sufficient information is available, however, to indicate that the distribution of costs within its annual budget for exhibits, conservation, salaries, and purchases of new material closely parallels that of other European military museums. In the matter of new acquisitions, this museum has been extremely fortunate in finding collectors and other benefactors who are willing to donate significant objects. Inasmuch as the museum does not charge for admission, there is no revenue obtained from this potential source. It likewise receives no profit for its own use from the sale of postcards, brochures, and small periodicals. Instead, the money collected is handled pretty much as a revolving fund and is used to print new publications.

The Musée Royal de l'Armée enjoys a fine reputation in Belgium and throughout Europe. This serves to attract a considerable number of visitors which presently averages 120,000 a year. The museum's location, some distance from the center of the city and the major tourist attractions of Brussels, does not work to its disadvantage, for many visitors who come to the famed Cinquantenaire Arch also stop off to see the Army Museum.

LEGER EN WAPENMUSEUM
(GENERAAL HOEFER)
Leiden, The Netherlands

The Netherlands' Leger en Wapenmuseum first opened in 1913 as an artillery exhibit. In 1928 its collection was expanded to encompass that of a regular weapons museum. From its beginning until 1945, the museum was located at the Doorwerth Castle near Arnhem. During World War II the Doorwerth Castle was destroyed and much of the collection was lost through bombings and pillage by the Germans. When the war concluded, the Netherlands Ministry of War decided that the Doorwerth Castle was not worth rebuilding and quarters for the museum should be sought elsewhere. Two buildings in the city of Leiden were founded which appeared to be suitable for housing a collection of military objects. The Netherlands Government leased these buildings from the city of Leiden, and what remained of the collection was moved to the new location.

Although the museum has been at its present site since 1945, it has been open to the public only since 1956. In the years since World War II, the present director has concentrated his efforts on locating objects lost during the war and acquiring new materials representative of all periods of Dutch military history. His problem has been a great one, and only now does he feel he is able to provide an adequate museum for the public. Many things in the museum reflect its newness. Some exhibits are incomplete, and part of one building is as yet unopen to public view.

In concept, the Leger en Wapenmuseum is not a national armed-services museum. It restricts its displays to those objects which have been used by the Dutch Army since about the mid-17th century. It has no naval specimens, for these are located in a special museum in Amsterdam, for many years the center of Dutch naval activity. It also does very little with airpower and limits its exhibits to a relatively few aircraft models. The museum thus fulfils the description in its title, for it is truly a weapons collection.

The museum collections are housed in two adjacent buildings. The main building is a two-story, rectangular-shaped structure containing some exhibits, the offices, workshops, and library. The second building is a single-story, square-shaped structure erected around a large open court. It is devoted entirely to exhibits, arranged in chronological order. The visitor first enters a hall containing military specimens dating from the mid-15th to the mid-18th century. Each succeeding hall around the square deals with a later and significant period in Dutch military history, so that the final hall visited displays objects currently in use by the Dutch Army. The open courtyard is the artillery park. It is lined with cannons dating from the 15th century to the more recent past.

The museum buildings were constructed about 1700 and obviously were not designed to be occupied by a museum. Nevertheless, they have been readily adaptable to the display of military collections. There has been some renovation, restricted largely to repair of the building. The rooms are fairly spacious and walls did not have to be removed to make space for the collections. Because the buildings are quite old they provide a very appropriate setting for displaying some earlier weapons and equipment, in particular, suits of armor, spears, pikes, old firearms, pistols, and swords.

The relative newness of the museum is reflected principally in its exhibits, for many of the objects on display need additional study before they can be adequately labeled or described historically. Since assuming his duties right after World War II, the present director has followed a systematic plan in readying the museum for the public. First, he had the buildings reconditioned and readied for the exhibits. The second step was to prepare the objects for display. This proved to be a rather monumental task because many of the acquired pieces had to be fully restored. Once the conservation and restoration were completed, the objects had to be studied. This is now the major undertaking at the museum and is far from complete. As a consequence, many objects are on display without any identifying data, and the visitor must temporarily be contented to examine these artifacts without the benefit of reading the essential information about them. The director recognizes this deficiency but says that he is not willing to label things until the needed study has been completed and he is certain he can place on the labels information which is absolutely correct.

The museum director approaches problems of exhibition from the viewpoint of the military historian. While he recognizes the validity of topical halls and has developed a number of them in the museum, he feels that the best form of display is to bring together all weapons, armor, uniforms, flags, and colors of a contemporary period. By so doing it is possible to present an integrated picture of the period. The director's objective is rather effectively fulfilled in his chronological halls, although there are still some obvious gaps in the material.

The zealous collection efforts of the director are evident in the great quantity of material he has been able to acquire. The present inventory includes slightly over 1,000 uniforms, 1,400 insignia and medals, 3,300 small arms, and more than 4,000 swords and other edge weapons. There are also over 200 guns and mobile artillery pieces. Added to this are about 12,000 specimens which are still packed and not yet properly inventoried. The museum boasts that it possesses the largest collection of powder testers and Dutch manufactured firearms in the world and one of the best collections of small swords in Europe.

The museum buildings are sufficiently large to house most of the materials without too much crowding. It is fortunate in having an artillery park with an area of 6,500 square feet, permitting the display of most of the heavy ordnance outside and thus leaving the exhibit halls relatively free of large pieces. The interior of the buildings contains approximately 70,000 square feet of potential exhibition space, all of which is not yet in use. This space appears to be entirely adequate for the items now on exhibit. The floor space is not crowded with a large number of show cases, guns, and pieces of artillery. The viewer has ample opportunity to move freely about the exhibits and examine all sides of any object. In most instances the specimens are displayed very attractively. There is a fairly efficient use of space, and cases are positioned to make the best use of both space and lighting. Much use is made of natural light, for the buildings have a large number of windows, and on sunny days the rooms are pleasant. However, when it is cloudy, generous amounts of artificial lighting are required to prevent the halls from becoming somewhat drab and shadowed.

The museum displays the bulk of its restored specimens, but there is a sufficient number of variations in individual collections to hold the viewer's interest. The director refuses to exhibit every piece in a given category because he believes the viewer should see only the best objects available. He will also remove an item on display if he acquires a replacement which is more attractive.

At present, the visitor cannot rely upon direction signs to permit him to make an orderly progression through the exhibits. However, when he enters the museum he is encouraged to buy a guide sheet, which is available at a nominal cost. This brochure contains a diagram of the museum, and each hall is briefly described as to content and period of time covered. A few members of the staff are also available to assist the visitor, and he may request help, both in locating various objects and acquiring information about them. The printed leaflet is especially useful because so many of the objects are inadequately labeled or have no descriptive data at all. The museum director feels that it is better, and indeed more intellectually honest, to give the viewer this kind of personal assistance than to let him read labels whose veracity may be questioned.

Because the museum's staff is relatively small and most of its time is consumed in research and preparing objects for exhibition, it is unable to provide more than a minimum of services to the public. The museum has a library of some 6,000 volumes, and the professional members of the staff are always ready and willing to assist students when they come to study. The staff seeks to answer as completely as possible all questions which are raised by visitors or are addressed to the museum by letter at the rate of about 2,000 a year. The museum is also prepared to offer its services in identifying specimens submitted by private collectors or interested people. Such other professional assistance as is deemed appropriate is also given the amateur collector, but the museum's staff resists any temptation to place a value on the objects brought to it for examination. The museum is unable to provide public lectures or show movies because it has no facilities for these services.

As a government agency, the Leger en Wapenmuseum depends for most of its financial support upon public funds, and other government agencies perform many services for it. The Ministry of War handles its budgetary requirements, and all its employees are in the Civil Service. However, in many respects it functions with the freedom of a private foundation. Bureaucratic interference is at a minimum, and it depends for much of its support upon an association of private citizens.

The museum has a board of 11 persons to provide general supervision for its operation. Only two board members are appointed by the minister of war, and one of these serves as the museum treasurer. The others are private citizens, all of whom take a keen interest in museum matters. In many respects the board is a self-perpetuating organization for, apart from the two members appointed by the minister of war, all other vacancies are filled by the board members themselves. Whenever a vacancy occurs, an effort is made to obtain the services of a distinguished man who has demonstrated great interest in the welfare of the museum. Once appointed, the members may serve as long as they desire. At the present time the museum director is also a member. The board exercises fairly loose control over museum operations. It reviews the annual budget and the museum treasurer controls the expenditure of funds. Beyond this the board does little else than offer suggestions to one of its members, the museum director. The director claims that he is in a most fortuitous position, for the board trusts him one hundred percent. It meets only once a year to transact business, tour the museum, offer suggestions to the director, and hold its annual dinner.

The museum is administratively organized around the work of three major departments. One is responsible for all weapons which date prior to 1840; another department cares for those arms dating since that year; a third handles all uniforms, colors, decorations, and paintings. The service sections, including the shops, are under the immediate supervision of the museum director.

At full strength the museum has a staff of 21 employees. In addition to the director, it has three professional positions, for

Le Musée Royal de l'Armée et d'Histoire Militaire, Brussels. A general view of the hall displaying small arms, edge weapons, and some artillery.

Le Musée Royal de l'Armée et d'Histoire Militaire, Brussels. An exhibit featuring the personal mementos of King Leopold II located in the Museum's Gallery of History (1831–1914).

Leger-en Wapenmuseum, Leiden. General view of the building.

Leger-en Wapenmuseum, Leiden. The hall displaying arms used by the Dutch Army in 1940 in the defense of the Netherlands against the German invasion

each major department is entitled to have its conservator or curator. The museum currently has only one employee with the rank of conservator. One of the departments is headed by an adjunct conservator, or assistant curator. The present head of the third department does not have a professional rank. The several technicians employed include one specialist in restoring firearms and swords, a specialist who restores paintings, a carpenter, and a painter. There are two full-time guards who conduct visitors through the museum and two men employed to clean the arms on display and perform other maintenance tasks. The museum also has two secretaries. The remainder of the employees are classified either as janitors or handymen.

The Leger en Wapenmuseum has achieved its present high level of attainment in part through the fine support given by the Dutch Government, which has greatly assisted by allocating funds for renovating the newly acquired buildings in Leiden and completing their adaptation for museum use. Of course, since the museum is a government agency, its operating costs are also paid by public funds. There is also a nominal admission charge, but the entrance fees net such a small return that they constitute no actual source of revenue. Hence, with the exception of certain philanthropic assistance, the museum is financed entirely by appropriation of public funds.

In common with a number of other military museums in Europe, the Leger en Wapenmuseum receives strong citizen support from an organization known as the Association of Friends for the Museum. The association performs a variety of useful public relations and other services for the museum, chief of which perhaps is the contribution the members make to a general fund used to purchase new objects for the museum. Through the generosity of the association, the director of the Leger en Wapenmuseum has been able to enhance the quality of his collection through the purchase of many significant pieces.

The director reports that the museum is on excellent terms with the Dutch Army. It automatically receives a copy of each new arm adopted by the army and acquires a complete new uniform whenever the army makes any changes. The army also willingly responds to requests for any other type of assistance which the director may send to it.

Compared to other military museums, the museum at Leiden has relatively few visitors. In part this may be the case because the museum has been open for such a short time and is now in an entirely new location. As knowledge of the museum spreads throughout the country, the number of visitors may well increase above the present 10,000 to 12,000 per year. The museum is also a bit difficult to find unless the visitor is a native of Leiden. Although it is not far from the center of the city and is only a matter of a few minutes' walk from the railroad station, it is completely separated from most of the other tourist attractions in Leiden, and directions must often be obtained before it can be found.

MUSÉE DE LA MARINE
Paris, France

The very excellent Musée de la Marine, which now occupies one wing of the Palais de Chaillot in Paris, ranks with the National Maritime Museum at Greenwich and the Sjohistoriska Museum at Stockholm as one of the finest naval museums in Europe. Handsomely quartered in one of Paris's newer and best-known buildings, this museum is the culmination of much effort by the French Government and interested private citizens to give suitable commemoration to French naval achievements. The museum was opened to the public in its present location for the first time in June 1943, during the Nazi occupation. At that time it was disguised somewhat as an exhibit of naval paintings to avoid undue difficulties with occupation authorities. The assembling of its collections, however, began many decades earlier, and, of course, much has been done since World War II and the liberation to achieve the objectives originally established for it.

French archives reveal that as early as 1679 Colbert, the famed finance minister of Louis XIV, recognized the value of preserving certain navy objects, particularly ship models, and issued instructions to the manager of the arsenal to give certain items proper safekeeping. In later years nobles of the French court kept their own "cabinets of curiosities" and gathered together many fine ship models, nautical engines, navigation instruments, maps, pictures, and engravings. In 1758, Duhamel du Monceau, the inspector general of the navy, offered his superb naval collection to the King. This extensive donation was duly installed in the Louvre near the Royal Library. Thus, the first Musée de la Marine came into existence, but as a special hall in the Louvre. A separate naval museum opened in 1801, but much of the material remained in the Louvre. This museum was reorganized some years later and was reopened to the public in 1833. Apparently most navy items were still retained as a part of the Musée de la Marine du Louvre and remained there until 1937, when French authorities decided to establish a separate museum at the Palais de Chaillot. World War II intervened before the new exhibit could be readied, and the opening was postponed until 1943.

In objectives, methods of display, and type of subject matter, the Musée de la Marine has very much in common with the National Maritime Museum in Greenwich and the Sjohistoriska Museum in Stockholm. All these museums expand their coverage beyond the navy itself to include other areas of marine activity considered of national significance. The Musée de la Marine, therefore, illustrates, through its exhibits, not only the French Navy but also the Merchant Marine, fishing fleets, yachting, and pleasure craft. The museum likewise attempts to have all aspects of naval affairs represented. Hence, it displays much naval art, examples of scientific achievement, the weapons of naval warfare and many illustrations depicting the history of French seamanship.

In one of its brochures, the museum draws attention to its patriotic theme by stating that it is dedicated to adventure and glory and thus endeavors to make a special appeal to the young. The brochure hastens to add, however, that the museum contains something for everyone and points to the permanent exhibition dealing with sport craft which is particularly interesting. Perhaps this is one aspect of marine activity one would hardly expect to find in a museum which is presumably concerned, to a great degree, with military history. Nevertheless, the navy museum at the Palais de Chaillot features a rather comprehensive display of sport activity ranging in subject matter from international races for sea-going yachts to underwater exploration.

It may be useful to speculate how far afield a museum might profitably go from its central theme of military history. There are, of course, no hard and fast rules which can be applied, because the incidents of military history do not transpire apart from a whole series of interrelated matters. Rather, it occurs within the context of fast-moving political, economic, and social events happening simultaneously, all of which have a far-reaching impact upon its course. In giving adequate interpretation of the events of military history, the curators of a military museum might find it useful to treat certain related nonmilitary subjects. The history of modern war is one of constantly expanding participation by masses of people and the channeling of more and more national resources to support military operations. How much of this should be shown? The military curator must make some difficult choices, and he must also resist a natural tendency to include so much that the central theme will be difficult to locate amid many related but side issues. Most military museums do not depart too much from their primary subject matter. Indeed, some fail to use their materials as meaningfully as they might in giving their displays a historical context.

The exhibits of the Musée de la Marine are almost equally divided between the military aspects of French sea power and other important national ventures on the seas. The exploits of the navy are recorded in the displays which are seen first by the visitor, but he leaves after examining exhibits dealing with French accomplishments in nautical science, exploration and navigation, the art of ship construction, yachting, racing, fishing, and pleasure craft. By having such a broad range of subject matter, the museum is able to introduce much variety into its exhibits and capitalize on the display of many types of ship models. It also can draw upon the interests of many groups and thus obtain a larger patronage than can a museum which restricts its exhibits principally to military objects.

Those familiar with the Palais de Chaillot will recall that it has two striking symmetrically curved wings and is located near the bank of the Seine River just opposite the Eiffel Tower. The exhibits of the Musée de la Marine are in the right wing of the Palais on the main floor. The offices, workshops, and other service facilities are located on the ground floor below. The wing containing the exhibits is divided into two parallel galleries running its full length. When entering, the visitor is encouraged to purchase a catalogue, for individual objects are numbered and reference to the catalogue is necessary either for identification or description. In this respect, the museum operates much as an art gallery. Very few other military museums in Europe follow such a system. The catalogue also directs the visitor through the museum to insure that he first makes a chronological pilgrimage through French naval history and returns to the starting point through a series of topical halls dealing with the rest of French marine activity. The two main galleries are not divided into rooms, but the exhibits have been prepared in well-defined units. This has been accomplished by placing cases or temporary wall partitions at right angles to the side walls.

The exhibits of French naval history are divided according to centuries or other particularly well-known historical periods. Almost the first thing the viewer sees is a model of Columbus's flagship the *Santa Maria*. Thus, the museum begins its presentation with the late 15th century. The 16th and 17th centuries are disposed of in a comparatively small space. This is not surprising, for the preservation of naval artifacts did not begin in earnest until after the middle of the 17th century. By contrast, about one-quarter of the first gallery is used to display objects from the 18th century. French naval power also achieved some of its greatest glory in this period. One feature of the 18th-century rooms is a set of excellent paintings of French seaports by the artist Vernet. From this point on, individual exhibits deal with such periods as the Revolution, the First Empire, the reign of Louis-Philippe, the Second Empire, and the Third Republic. The presentation of French naval history concludes with a large display of objects which date from the beginning of World War I and extend to the present time. The remaining portions of the museum are also of great interest, and the exhibits are done in excellent taste. However, they are not of major concern to this study.

Because the Musée de la Marine is comparatively new and modern museum techniques are used in developing its exhibits, this museum undoubtedly creates a very favorable impression upon most viewers. The long, well-lighted galleries appear quite formidable and are very attractive. Care has been taken to see that worthy artifacts are treated as objects of fine art and thus displayed to best advantage. The museum contains many treasures, some of which were presented as gifts to French sovereigns and heads of state. These are well displayed. Every naval museum boasts of its fine collection of ship models. Certainly the Musée de la Marine has its share, and those selected for display represent the best work of skilled craftsmen. Objects such as these should be carefully examined if their details are to be fully appreciated. This is time consuming but most worthwhile. Since there is so much of value to see, several trips to the museum may be required. A brochure suggests that the viewer not attempt to see everything at once. He is advised to "take a quick look around, so as not to get too tired, and then to come back again to look closer at the things one finds the most interesting."

There are a great many objects on display, but the museum does not appear to be overcrowded. In addition to the many ship models which are found in almost every exhibit, there are two rather famous craft on display which were actually used on state occasions. One of these is the charming small boat belonging to Queen Marie-Antoinette. The other is the long white barge adorned with gold used by Napoleon on his entrance into the harbor of Antwerp and afterward, under the Second Empire, to carry the Empress Eugenie when she visited Brest. Both of these objects are placed in the center of the hall, where they can be carefully examined on all sides. There is plenty of room in every exhibit area to accommodate a large number of visitors, although in most instances no impression of spaciousness is created.

In preparing its exhibits for selected periods in French naval history, the Musée de la Marine follows the pattern of most European military museums. It combines weapons, some heraldry, documents, models, pictures of naval engagements, and portraits or busts of celebrated naval commanders to create an over-all impression of an age. Dioramas are not used. The progression of the shipbuilder's art and with it the change in the character of the French Navy are well illustrated by ship models as one goes through the galleries.

No effort is made to tell in detail the history of any war, century, or other period. Great naval battles in which France fought come to the viewer's attention through the paintings or objects associated with them. The weapons, type of ships, and other equipment used by the French Navy are revealed also, but the exhibits lack much useful background data which interpret the navy's role as an extension of French power in international politics. The casual viewer will see a lot of interesting objects which may delight his fancy, but he may emerge from his visit with very little additional information about French naval history. This deficiency in contextual historical interpretation within the exhibits themselves is offset somewhat by the explanatory material contained in the catalogue. Objects are all identified in the catalogue, and some descriptive information is added in many instances. Rather extensive biographical data are provided for the most celebrated naval commanders. If one takes the time to read all this material, he will gain a considerably deeper insight into French history.

Very little attention is given in this museum's displays to the rank and file of the French Navy. The claim is made that the museum's treasures remind the visitor of the sailors and ships of past times. However, the whole museum has been fashioned to recall the fame of departed heroes as well as the wonders of the shipbuilder's art. In general, the anonymous French sailor has been lost in the process.

At present the policy of the museum is to maintain permanent exhibits. From time to time these are augmented with new material. Some objects were damaged during World War II, and the staff is now devoting as much time as possible to restoring and repairing these. No space has been set aside for special exhibits on the main floor, but the museum program carries a sufficient number of special features to guarantee a considerable degree of constant public interest in its activities.

The inventory of the Musée de la Marine contains about 7,850 objects, a goodly portion of which have no military connection. Surprisingly, the museum has very few uniforms. The stock includes only 80 complete uniforms or pieces of uniforms, 35 hats, and 60 uniform accessories. The museum possesses about 2,200 medals and insignia, 215 small arms, 120 swords and other edge weapons, and 45 large guns or cannon. It likewise has 7 full-size small boats and yachts. The museum's military models are some of the most important items in its collections and include 410 ships, 25 aircraft, and 370 items of ordnance. In addition to the objects on display, the museum possesses rather extensive research files. Included are some 29,500 iconographic documents, 32,500 photographs, and a library of approximately 9,500 volumes.

The Musée de la Marine's two principal galleries provide it with 50,570 square feet of exhibit space. It has no exterior area for display, but the Palais de Chaillot is not particularly designed to accommodate outside exhibits. Considering the present manner in which the museum uses the total area of the Palais allotted to it, it is doubtful that much additional room could be found for more displays. In reporting its space utilization, the museum makes no distinction between the area housing its reference collections and that given over to dead storage. Rather, it notes that a total of 7,500 square feet is set aside for all forms of storage. Laboratories and workshops occupy 7,000 square feet of area, offices about 1,000 square feet, the library 2,150 square feet; and a salesroom located at the entrance to the exhibit galleries occupies a space of approximately 3,200 square feet. The museum also has a fairly large receiving area, containing slightly over 2,000 square feet.

Although the Musée de la Marine has a relatively small staff, it conducts a most active program aimed at increasing public interest and participation in its work. Fortunately, the staff does not have to depend fully on its own resources, for in its program it is ably assisted by a strong association of "Friends of the Naval Museum." Because the staff is presently concentrating upon the restoration of the items in its inventory, it must restrict its research activities to ship models now under construction, acquiring data on some objects in its collections, and the preparation of formal replies to questions received from other museums. No extensive independent research program is known to be contemplated by the curators in the near future. The staff members identify military objects or provide other information when requested to do so. However, they do not feel that they are adequately set up to do much of this type work. Hence, the service is performed very informally, and no estimate is made of the number of inquiries handled annually.

In addition to the Musée de la Marine, there are five small navy museums in France. Only two of these are now open to the public. Because these museums are rather limited in their own resources, they look to the Musée de la Marine for some assistance. Their requests for help are honored, and they receive needed data, assistance in restoration of objects, and valuable advice.

The Association des Amis des Musées de la Marine is one of the most active organizations of its type in Europe. Its fundamental concern is the welfare of the museum, but its interests and programs go beyond the institution to embrace many matters related to the French Navy and other maritime activity. The museum is essentially the center through which this organization functions. The association was established under the patronage of the Minister of Marine (Navy) and has as its stated objectives "the coordination of researches into maritime history and of all matters relating to the Navy." In accomplishing its objective in the field of historical research, the Friends of the Naval Museum publishes two illustrated quarterly reviews. One is a general naval magazine called *Neptunia,* and the other, *Triton,* also devotes attention to all matters pertaining to the sea but specializes in ship models. The latter review is a boon to those who have the hobby of building ship models. Both magazines enjoy an excellent reputation among marine enthusiasts and provide suitable vehicles for publishing the research of naval historians and writers as well as the work of the Musée de la Marine staff.

Apart from its support of research in maritime history as revealed in its quarterly publications, the association performs a number of services at the museum for the public. At least once a month the members of the group meet in one of the exhibition halls for a major lecture. They may also attend a special meeting once a week. At this meeting a film is often shown or they go to one of the exhibit areas for a special presentation. The public is invited to some of these events. The association also provides the museum with a guide service. Tours for small groups can be arranged, but advance notice must be given. A member of the Association will then be present to escort the party through the museum, discuss the various exhibits and answer the inevitable questions. One final service the association performs is the operation of the museum salesroom. Many books and publications as well as souvenirs are available for purchase. The association receives all profits realized from the sale of these items.

The immense value of this organization to the museum is

readily apparent. Because it has such an ambitious program and engages in activities which are both intellectually stimulating and entertaining, it can help to attract rather broad public support for the museum. It is likewise a source of encouragement to the staff, for it can render assistance where needed. This aid also includes some financial support, for part of the Association's funds is used to purchase objects for the museum collections. A number of other artifacts are probably donated to the Musée de la Marine through the efforts of the Association.

Until 1919, the navy collection in the Louvre was administered by the director of national museums. It was then transferred to the control of the Ministry of Marine, the department which maintains jurisdiction over the Musée de la Marine. The method of official control over museum operations is fairly similar to that found elsewhere in Europe, although the Navy Ministry demonstrates a greater degree of interest in the museum's activities and welfare than is often the case in similar instances elsewhere. This interest is shown through normal budgetary review, the composition of the museum's professional staff, the selection of the board of directors, and the official patronage accorded the "Friends of the Naval Museum." Through all these channels, the Ministry of Marine is able to provide general guidance to museum affairs, while permitting routine operations to be the province of the director and his fellow curators. The navy itself has no official relationship to the Musée de la Marine. It donates any objects of interest to the museum which are no longer in use, but it performs no other services.

The board of directors has seven members, all of whom are appointed by the Minister of Marine. The chairman of the board is always equivalent in rank to an admiral. At present the chairman is the paymaster general of the navy. The remaining members are all chosen from the various major sections of the Ministry. As is the case with so many other museum boards, the members are not required to possess any experience in museum management prior to selection, although they may have shown some interest in the enterprise at one time or another. Their qualifications are thus rather indeterminate, and often a member may be given an on-the-spot appointment when a vacancy occurs. The duties of membership are not too arduous and are performed apart from regular professional responsibilities. The board in general restricts its attention to matters of major policy. It approves the museum budget and makes its recommendations to the Ministry of Marine. All items of major repair to the facilities are cleared by the board, as are purchases of expensive objects and the lending of any portions of the museum collections. The board is able to keep in fairly close touch with the general activities of the Musée de la Marine because it meets once a month. This is considerably above the average for military museums.

The director of the Musée de la Marine undoubtedly enjoys the confidence of the Ministry and the board, for he has been in this position since the museum opened to the public in 1943. He and the other two members of the professional staff are career officers in the French Navy. Their assignment to the museum is a permanent one and is made in recognition of talents these officers possess in the fields of maritime history and museology. The permanent assignment of professional officers to work in a military museum is very much in the European tradition. Their other requirements as career officers are waived for the most part or made subordinate to their primary career as military museollogists. The appointment of these professional naval officers to manage the Musée de la Marine does guarantee, to some degree, that the museum will be in the hands of ardent marine enthusiasts or at least persons who fully appreciate the historical traditions of French sea power. Further, their indefinite period of service in the museum gives them some sanctuary from possible service-type pressures and permits them to develop a viable institution in accordance with the lines of policy set by the board of directors.

Administratively the museum is organized into five departments, which correspond to the essential functions performed. Division according to subject matter has been rejected as a practical plan of organization. The departments are listed as the library; iconography, which cares for the museum's extensive collection of pictures and engravings; exhibits; conservation; and maintenance. Perhaps this type of organization is best suited for a museum which has so few on its professional staff. The director is thus able to keep his span of control quite limited and concentrate into one department the full range of subject matter covered by the museum's exhibits.

The Musée de la Marine presently employs 48 people, which is about its normal complement. Only three of these perform duties at the curatorial level; they are the professional naval officers on the staff. The director and subdirector both have the rank of capitaine de fregate. The third officer is rated as an officer de equipage and supervises many of the administrative functions at the museum. The remaining members of the staff are technicians, specialists, clerks, or guards. Five persons are assigned to work in the library and catalogue the collections, and three look after the photographs and engravings. About 20 are employed in the various workshops in such specialized capacities as modelmakers, painters, cabinetmakers, and weapons conservators. The director's office has two secretaries or clerks, and the museum employs eight guards who also are required to keep the exhibit halls neat during visiting hours. Janitorial service as well as building repairs is handled by private contract.

The finances of the Musée de la Marine appear to be handled in a relatively uncomplicated manner. All operating funds are obtained by appropriation in support of the budget approved by the Ministry of Marine, and upkeep of the property is a government function. Since the Palais de Chaillot is a major tourist attraction and one of the more impressive buildings in Paris, this structure is in a fine state of repair. The excellent condition of the museum bespeaks great pride in its appearance as well as its work, and sufficient funds are apparently always available to keep it at its present high state of attractiveness. The additional assistance given to the museum by the Association des Amis de Musée de la Marine through the purchase of fine objects for its collections and in providing other services cannot be fully evaluated in financial terms. Nevertheless, such assistance contributes much toward raising the quality of this institution. The museum also charges a nominal fee for admittance, but the revenue goes to the Treasury.

In its effort to stimulate some patriotic fervor in those who view its exhibits, the Musée de la Marine achieves considerable success. It capitalizes well upon the outstanding accomplishments of France in shipbuilding, exploration, and the development of world trade. It recalls with attractive displays the glorious chapters of French naval history which were written when the nation ranked as one of the great world sea powers. Those who come to visit its fine exhibits, and about 150,000 do every year, provide themselves with a most enjoyable venture in one of the best naval museums in Europe.

MUSÉE DE L'ARMÉE
Paris, France

For many people Napoleon Bonaparte is the symbol of French military power. Certainly under his leadership France attained a pinnacle of strength unequaled since his time, for he fashioned the French citizen army into the dominant military force in Europe. In earlier days, also, France was consistently one of the major powers on the continent, and her history records many great military and naval achievements. But no hero emerged to captivate the imagination of his nation as did Napoleon, nor did any figure put his stamp so completely upon an era in French history. His military accomplishments are fittingly commemorated in his impressive tomb and in the exhibits of the Musée de l'Armée, both of which are located at the historic Hôtel des Invalides in Paris.

Louis XIV founded the Hôtel des Invalides in 1670 as a home for old or disabled veterans and had it constructed in what is now the central section of Paris. It is an immense edifice of many courts and wings and has been described as "grandiose in plan but commonplace in detail." The structure is dominated by the domed portion of the building which houses Napoleon's tomb. Behind this is the church, Église St. Louis, and the great rectangular court which contains the Army Museum. The pensioners of the French Army have left, for the most part, and the Hôtel des Invalides has become in later years one of the principal tourist attractions of the French capital.

For many years the ground floor of the great court of the Hôtel des Invalides has housed military collections, but the present Musée de l'Armée has existed as an entity only since 1905. Previously, separate wings displayed the collections of two distinct museums. One was the then unsurpassed National Museum of Artillery, which, as the name implies, was strictly an arms and armor exhibit. The other was known as the Historic Museum, but its displays were chiefly military in character and devoted to many activities of the Ministry of War. The Musée de l'Armée was created when these two collections were combined within a single jurisdiction.

The building is not particularly ideal for a museum, but it is large, and there is apparently sufficient space to display most of what the staff desires. The portion of the Hôtel des Invalides in which the Musée de l'Armée is located is a two-story rectangular barracks with a veranda along all sides facing the inside court. The exhibition rooms are fairly good sized and have high ceilings as befit a barracks constructed many years ago. Some renovation has been necessary to make them usable for museum purposes, but there has been no significant alteration to the building's basic plan. Since all the rooms are not interconnected it is impossible to begin at one point and have a complete tour of the museum by going directly from one room to the other. Instead, different groups of exhibit halls are separately entered from the veranda, which is quite wide and provides a considerable protected space for the exterior display of a large quantity of cannon and other sizable pieces of ordnance.

For the most part the Musée de l'Armée's exhibits are topical, and no effort is made to provide a comprehensive description of French military history. The special halls which display Napoleonic relics have a good many historical references, as also do the halls exhibiting the memorabilia of World Wars I and II. But this is about the extent to which excursions are made into chapters of French military history. The domination of Napoleon at the Hôtel des Invalides is evident from the two halls given over to his personal mementos. One is devoted principally to the period of his life when he ruled France and led its armies on their celebrated campaigns throughout Europe. Hence, most of the objects on display are military and include Napoleon's uniforms, his personal weapons, decorations, and other equipment. Other materials are also included which give a fairly good description of Napoleon's armies, although there is no detailed discussion of the military action of the period. The second Napoleonic hall is largely nonmilitary in character, for it includes family and personal items ranging from dishes and clothing to the bed in which Napoleon died. Such objects normally are not found in a military museum, but the location of the Musée de l'Armée and the attention given to Napoleon make their inclusion somewhat logical and expected. The addition of these Napoleonic relics to the military collections adds somewhat to the interest of the Musée de l'Armée and no doubt brings additional visitors.

The remaining exhibit halls of the museum follow the usual pattern for a display of arms and armor. Attention to ancient times is given through an interesting exhibit of excavated weapons which include some Roman items and those from prior periods. The museum also has one of the best collections of armor in Europe, and two halls have been prepared for their display. One was opened several years ago and is perhaps the finest hall in the institution. Good exhibition techniques are employed and the armor is shown to excellent advantage. The museum director has obvious pride in his new hall and points to it as an example of the museum's progress toward modernization. Other rooms display artillery, small arms, uniforms, and memorabilia of the French Air Force in World War I. Very little has yet been done with the materials from World War II, but the museum has plans to build new exhibits as soon as some additional space can be made available. The Musée de l'Armée also has a rather elaborate display of model soldiers of all nations. These are lined up in processions, and there are some battle formations.

Because access to the various halls is from a veranda which also displays many cannon, the visitor may feel that the veranda itself is one long exhibit hall. The museum is indeed fortunate in having such a fairly extensive exterior area which is partially protected from the elements. The student of artillery may encounter some difficulty in making a close examination of many of the cannon. Some of the museum's finest pieces are unpended and placed against the wall. They cannot be seen in all details or photographed without flash equipment because the veranda is often quite dark. However, this form of display does permit the placing of a considerable number of items in a space that must also be used as a passageway.

The total number of objects in the Musée de l'Armée's inventory is a bit difficult to calculate, but the museum is one of the largest in Europe and has one of the most varied collections. It has 300 complete uniforms and over 3,000 pieces of uniforms and personal equipment. Added to this are 4,630 medals and insignia. Most of the uniforms, parts of uniforms, decorations, and medals were donated by private families and are personal souvenirs once worn by French officers. The small-arms collection consists of 1,580 rifles and 450 pistols, and its edge weapons include 930 European swords and sabers and 380 from oriental nations. Its heavy ordnance numbers 430 cannon, a few of which are complete with carriage equipment. There are also 350 artillery models on display. Armor comprises a considerable portion of the inventory and includes 475 suits and separate pieces of armor dating from the 15th to the 18th century, 340 pieces of equipment for men and horses, 290 pieces of headgear, 90 shields, and 1,300 shafted weapons such as halberds and battle axes. Added to this is a smaller oriental armor collection of 46 suits of armor and 300 halberds. The museum also can offer assistance to the researcher through use of its 20,000-volume library.

In the amount of space allocated for display, the Musée de l'Armée is similar in size to the Leger en Wapenmuseum at Leiden and the Tøjhusmuseet at Copenhagen. At present this amounts to 69,400 square feet and includes both the individual rooms and some of the interior corridors and vestibules in which objects are displayed. No accurate estimate can be made of the actual space used for outside exhibition on the veranda of the building, because the veranda functions principally as an access route to the museum halls, and objects are placed intermittently along the passage or against its walls. The veranda itself, however, contains about 11,000 square feet. The amount of storage space used by the museum is not known. Individual workshops have been provided to meet the museum's need for the preservation of armor, restoration of uniforms, painting, and carpentry, but their area is likewise not known to the writer. The museum also has a small exhibits laboratory. Its offices and other service rooms total about 8,100 square feet, and its library is housed in a space of 1,300 square feet.

As a public institution the Musée de l'Armée is part of the Ministry of War, but the relationship is a somewhat tenuous one. The Ministry has certain prerogatives of control which it exercises rather loosely. The museum has been operating somewhat autonomously for many years to the apparent satisfaction of the Ministry. Its future is well assured, because it functions as an instrument of patriotism and plays a somewhat supporting role to the suitable memorialization of a major national hero. In short, it is a going concern and appears quite able to function with a minimum of bureaucratic interference.

The Musée de l'Armée is unique among European military museums in not being financially supported by the government. Such support is apparently not necessary, because its expenses are fully covered by the admission fees charged the large number of visitors for entrance to the Tomb of Napoleon. If the museum were not located at the site of this celebrated tourist attraction, it is quite possible it might have to seek other sources of financial support. Even though the Ministry of War does not include the fiscal operations of the Musée de l'Armée in its own budget, it indirectly exercises the right of budget review because it must give approval to the decisions of the museum's policy-making body. Apart from its regular source of income, the museum receives some funds from its society of friends, but this does not amount to a great deal. Further, it is not responsible for the maintenance of its building; this duty falls to the Ministry of Fine Arts.

A board of directors serves as the museum's major governing body. Composed of 20 individuals, the membership is equally divided between representatives chosen from government agencies and those selected from the public. Of the 10 drawn from the ranks of government, 4 are employed within the Ministry of War, and the remainder come from the Ministry of Fine Arts, the Ministry of National Education, the Army, Treasury, and Court of Public Accounts. The 10 private citizens who serve are renowned historians or prominent military collectors. Included in the citizen group are two members of the National Assembly, both of whom also happen to be former army generals. Appointment to the board is made by the Minister of War, and the term of service is for 5 years. In part, the board is a self-perpetuating organization, for it suggests names of prospective members to the Minister of War who usually responds by appointing them.

The board of directors is moderately active, meeting on an average of four times a year. It gives its closest attention to preparing the budget, but it also develops, in general terms, the main lines of policy for the museum. It submits its decisions to the Minister of War for his approval, and this is usually forthcoming without delay. The present director of the Musée de l'Armée is most satisfied with this type of arrangement and with the composition of the board. He is assured of reasonable autonomy in conducting museum operations, and he can draw upon the support of men who are widely placed in government or who possess considerable public stature. The historians and professional men who serve help to give the museum a sound academic foundation and acceptance and are a ready source of valuable information, whereas the political functionaries are in a position to assist the museum in its intragovernmental relationships or in reducing the delays which often are encountered in obtaining needed government services.

The director of the Musée de l'Armée is a retired army general who has maintained an active interest in museum affairs for many years. As a professional soldier he was not trained specifically for a position of museum management, but he has deepened his insights through acquired experience. He administers an institution which does not place a major stress upon the interpretation of French military history. Rather, he emphasizes the museum's custodial and commemorative functions, as no doubt his predecessors also did. In exercising his responsibilities, he has a great deal of freedom and claims to enjoy the confidence of the Minister of War. His primary duties are, of course, at the museum, but he does perform some nonmuseum functions. For example, he occasionally lectures at the École Militaire on subjects related to museum activities or some phase of French military history. He also has one unusual responsibility which entails looking after the Église St. Louis at the Hôtel des Invalides. In explaining this unique task, he quickly points out that his duties are strictly temporal and are most concerned with supervising the premises and scheduling some of the church's activities.

Because the Musée de l'Armée is rather limited in its program and has a relatively small staff, the director keeps its administrative structure quite simple. There are no functional divisions or formal departments and, with the exception of the technicians in the workshops, the director gives immediate supervision to the museum employees. The curators on the staff do not have the opportunity to overspecialize, for each is placed in charge of several halls and periodically rotated in his assignment. One curator has the added duty of supervising the technicians. Most European military museums frown on this method of utilizing the professional staff, for they normally stress developing a staff of experts in particular fields. Usually, only the director of a museum is something of a generalist and most often he too is a specialist in armor, swords, uniforms, or some other particularized subject matter. How often the curators at the Musée de l'Armée change their assignments is not known, nor can the results of such a policy be adequately evaluated without further study. The director feels this system gives him the most efficient use of his present curatorial staff and offers each man some variety in his duties as well as a broader understanding of the museum's total work.

There are presently about 50 persons employed at the Musée de l'Armée, and approximately half of these are guards. A fairly good number of guards or warders is needed because many of the halls are not interconnected, and it is difficult for one guard to serve a very large area. The museum has only three curators, and they work principally with the exhibits. Only four technicians are now employed to service the collections and meet the museum's requirement for upkeep of its exhibits. These include a painter, a carpenter, a specialist in the preservation of uniforms, and a conservator of armor. Ten employees provide the museum with clerical and administrative help. Maintenance and janitorial personnel are not included among the regular employees, but the needed service is performed by contract. The Musée de l'Armée does not have a librarian at present. Reorganization of the library is a project to which the director is giving a great deal of attention. He hopes he can have a person skilled in languages to supervise the library's services, for he believes a linguist can be of great assistance to the arms specialist and the research student.

The Musée de l'Armée does not classify itself as a major research institution. The student may come to examine its collections and to conduct his personal research in the museum library. But the staff is too limited to perform little more than the routine tasks required for the adequate upkeep of the museum. None of the staff is now engaged in any special research projects, although one of the curators is a fairly regular contributor to a historical publication. Some research is required in connection with objects in the collections, and this is performed by the staff.

Such study is necessarily limited in scope. Like all military museums, the Musée de l'Armée receives a sizable number of questions from the public, although not so many as might be anticipated for such a large institution. These questions number about 600 a year and are answered, for the most part, by the director. He claims he cannot answer many of the more detailed questions, and so he turns to some of the historians on the board for assistance.

The museum undertakes no program of lectures or other presentations beyond an occasional Napoleonic exhibit which is arranged for by a private organization and offered in one of the museum's rooms. Its other principal service to the public is to offer guided tours through pre-arrangement; these are normally led by the curators and, on occasion, by the director. The museum does not lend any objects to the public, but it will lend its duplicates to some local museums. It likewise offers informal professional advice to other museums when requested.

Unlike the Musée de la Marine, which is quite closely connected with the navy and the Ministry of Marine, the Musée de l'Armée has a very tenuous relationship to the French Army. The army makes no direct effort to influence its activities, although its interest in the museum can be conveyed through its representative on the board of directors or through other members who represent the Ministry of War. The army does not donate objects to the museum, and the museum must depend entirely on donations from private sources or purchases to augment its collections. Infrequently, the director may call on the army for some logistic-type assistance. Officers studying in French military schools frequently visit the museum, and the staff informally places its services at their disposal.

The museum is supported by a reasonably active society of friends (Société des Amis du Musée de l'Armée). Some money and donations are received through its efforts, but the amount of either is not great. The organization sells publications and souvenirs at the museum, and uses the profits to finance a portion of its own activities. It likewise publishes a revue which features special articles on arms and armor in addition to some which deal with military history. It is the chief method used to publish research performed at the museum by the society's members and others. The director feels the organization of friends has been very helpful to the museum, particularly in publicizing its exhibitions.

A tour of the Musée de l'Armée cannot be completed quickly, for there is much to see. For the most part the museum is not modern, and some of the halls reflect the difficulties imposed upon the staff by an old building. Nevertheless, the visitor can find many excellent objects which are well exhibited. Over the years, the museum has developed the traditions of an arms and armor display rather than one providing a comprehensive coverage of military history. Imposed on this is a responsibility to give suitable commemoration to Napoleon. If one first visits the Tomb of the Emperor and follows this pilgrimage with a tour of the Napoleonic halls of the Musée de l'Armée, he may easily gain the impression that the museum is little more than part of a public monument to this celebrated hero. Should he look further, however, he will discover one of the largest collections of armor and swords in western Europe.

TØJHUSMUSEET
Copenhagen, Denmark

One of the finest and best displayed collections of arms in Europe is located in the Royal Arsenal Museum (Tøjhusmuseet) in Copenhagen, Denmark. This museum was founded in 1838 as "The Historical Collection of Weapons," but the arms exhibited came for the most part from the 17th century. The heart of the collections is the armories of the Danish Kings and the Ducal Houses of Gottort and Oldenburg. The arms are not entirely Danish but are international in scope and are drawn chiefly from western Europe. However, the collection of artillery is mainly Danish. The museum inventory includes both military and civil weapons, uniforms, and other equipment of warfare.

The Tøjhusmuseet cannot be considered a general armed-services museum or one devoted to any branch of the Danish armed forces. Rather, it is an arms museum and seeks only to exhibit the fine items in its collection. It makes no effort to tell the story of any particular period in Danish military history or that of any service. It possesses the essential characteristics of a national museum as revealed in the scope of its collection, its relationship to the Danish Government, and the position it occupies as the foremost of its type in Denmark. Its objectives as a museum are simply to display its excellent collections as attractively as possible and to place its entire resources at the disposal of the serious student.

The Tøjhusmuseet is housed in three ancient buildings in the heart of Copenhagen. The largest of the three contains the exhibits. The offices and workshops are in a second building, and the third one provides the museum with its storage space. The exhibition building was originally built as an arsenal by a Danish King around 1600. The storage building was constructed a few years later as the royal brewery and storehouse for grain. This building is little changed from its original state. As a result, it presents some difficulties in so far as its present purpose is concerned. As the ground floor is quite damp, it provides storage only for nonmetallic objects. The difficulty encountered in heating the building somewhat restricts its use during the winter. The upper floors are too hot in summer and too cold in winter. With such extremes of temperature these floors are little used. The museum's long-range plans call for the complete renovation of the storage building, but because of the cost the remodeling will not be completed for a number of years.

There have been few architectural changes to the exhibition building since it was converted to use as a museum. Because it was originally designed as an arsenal, the director of the museum feels that few major changes are really necessary to display military objects. The building is of considerable length, and neither the ground floor nor the second floor has been partitioned to make smaller halls. The ground floor is the cannon or artillery hall, and two rows of cannon stretch its entire length. Thus, the display gives the appearance of an arsenal, which is precisely what the museum's staff intends. The cannon are placed chronologically so that their evolution can be followed from the earliest to the present state of the art.

The displays of edge weapons, pistols, and other firearms are located on the second floor of the exhibition building. The walls of the rooms are lined with cases containing swords displayed vertically. Across a rather spacious aisle from the sword cases are others containing pistols and daggers. Lining the central aisle of the hall are open racks of guns and other weapons. These are the study collections which are readily accessible for close examination by the serious student and other interested viewers. A few colors and standards are also displayed in this hall. They are suspended from the ceiling so that the entire surface of the color can be observed. Part of the third floor of the exhibition building is also used for display space. Here a number of the uniforms of the Danish armed services together with some important insignia and medals are shown. The rest of the third floor is used for storage and contains the museum's rather extensive stock of uniforms.

The museum divides its inventory into three categories. First, its catalogue lists about 40,000 single objects which form the key portion of the collection. The second division contains about 67,000 duplicates, mainly Danish military weapons from the 18th, 19th, and 20th centuries. The third category is a rather considerable collection of falsified weapons which are used by the staff for comparative study. The first group contains some 800 uniforms, 3,400 parts of uniforms, 900 insignia and medals, about 4,300 small arms, 4,000 swords and other edge weapons, 700 heavy guns, 3 armored cars, 7 military aircraft, and about 18,000 plans or drawings that are associated with the ordnance contained in the museum. The inventory also lists a number of military models which include 25 aircraft and 250 ordnance.

The exhibition building has an allocation of about 71,000 square feet for displays. Since the ready reference materials are also housed in the same room with part of the exhibits, no separate space is set aside for such items. The museum has about 170,000 square feet of storage space at its disposal, but only a fairly small portion is used because the storage building is so inadequate. There is also a library of some 6,000 volumes occupying a room containing a little over 1,200 square feet. It is located in the same building as the offices. Office space is limited to 3,000 square feet, and laboratories and workshops add another 3,200 square feet to the museum facilities. There is also a lecture hall of about 2,000 square feet.

The method of exhibition adopted by the museum is entirely consistent with its character as an arsenal. Prime emphasis is placed upon the attractive display of weapons and these immediately capture the viewer's attention when he enters the exhibition hall. The arms are placed in straight and orderly rows running the length of the building and the viewer may easily gain the impression that he has entered an orderly and well-stocked arsenal. Signs direct the visitor through the halls in an effort to insure that he may acquire an orderly presentation of the exhibition. A guided tour is unnecessary for the average visitor, for directions are adequate and a careful reading of the labels will give him the essential data about the items he is viewing.

All objects within the museum are well displayed. There is no clutter; nothing is crowded. All pieces are in a state of excellent preservation, so that the objects themselves are esthetically attractive. However, certain limitations are imposed upon the opportunity for flexibility of display because the objects are exhibited apart from any context of military history, and the staff firmly adheres to its basic concept of giving the museum the general appearance of an arsenal. Any viewer who is fascinated by a large and diversified collection of weapons will find the Tøjhusmuseet exceedingly interesting. The upper two floors of the building are well lighted, but the ground floor with its heavy pillars and vaulted ceilings depends for the most part upon natural lighting. This is adequate, or fairly so, on sunny days, but on cloudy days the hall is quite dark.

As might be anticipated, the form of exhibition precludes the making of many changes to the major displays. The artillery or cannon hall and most of the exhibits on the second floor are considered permanent. New pieces are added to existing exhibits, but these are the only changes normally made. However, floor space is available for small, special exhibitions. These are held periodically and publicized to attract new visitors to the museum.

The director of the Tøjhusmuseet considers his institution one of the major military museums of Europe. Although a sizable increase in his budget is considered most desirable, the director does not permit pecuniary limitations to give the impression that his operation is a parsimonious one. He looks upon the museum as a service institution to the public. To that end he has staffed it with qualified people who are able to provide the museum with high quality exhibits, are able to render many useful services to the public, and have attained high professional recognition among their colleagues. As a result, the Tøjhusmuseet ranks very high

The Tøjhusmuseet, Copenhagen. A view of hall exhibiting small arms. The reference collections are also displayed in the center of the room.

The Tøjhusmuseet, Copenhagen. The cannon hall on the ground floor.

among other military museums in research and service to the public.

In addition to the routine research involved in identifying artifacts and their connection with military history, the professional staff engages in an extremely ambitious program of independent research. Apart from a number of public relations type publications, the museum produces a series of professional books under the general title of "Tøjhusmuseets Skrifter." This contains the volumes written by members of the staff and includes such articles as "Gunmakers from the Sixteenth Century," "Swords from the Middle Ages," and "History of Danish Powder Factories." Added to these publications are the catalogues and articles about the museum which appear in periodicals. Several of these have been written by the director. The staff also answers many inquiries a year from the public and is always ready to identify specimens or provide professional advice to amateur collectors. The museum budget allocates no separate funds for research, but this service is considered an important museum responsibility and is a duty which every member of the staff is expected to perform.

The staff performs several other services which should be mentioned. One of the most important of these is the comparatively heavy lecture schedule maintained by the professional staff of curators. An average of several hundred soldiers in the Danish Army and Cadets in the Officer Candidate School come to the museum every month to receive instruction in the general field of military history. As these lectures fall outside the regular activities of the museum staff, the individual lecturer receives an additional stipend for this service. The museum also provides guided tours for the general public. At least one tour is scheduled every day and two are conducted on Sunday. Because of the routine nature of these tours they are given by members of the conservation staff, i.e., the technicians. However, the special tours for high-school groups or university students are conducted by the museum curators. All the tours are free to the public. Further, the staff is always willing to assist any students who may come to use the museum, library, or study collections. Books and artifacts, however, are not lent to the public, but the museum will lend some of its collections to other museums and to military barracks. One other service provided by the Tøjhusmuseet is to assist a local museum in problems related to the restoration of specimens. It will actually do the job if the local museum is unable to provide this kind of service for itself.

The Tøjhusmuseet is a government agency and is included within the Ministry of Defense. The museum has no board of directors or any other body apart from the museum staff which is responsible for setting policy. The director's only superior officer is the Minister of Defense. Thus, he is virtually in complete control of the museum. He takes full initiative in all matters related to his organization. However, he must account monthly by a special report to the Ministry of Defense for the spending of all funds. He receives absolutely no guidance in return. The director confesses that his position is somewhat unique when compared to that of other military museum directors. He feels that the results are worth this type of managerial freedom, for he has excellent rapport with the Minister of Defense and is able to secure any services desired without major delay.

The museum is organized into six departments. These are: small arms; swords, daggers, and armor; artillery, aircraft, and engineering; uniforms, heraldry, colors, barracks, and instruments of punishment; preservation or conservation; and administration. With the exception of the last department, which is under the immediate supervision of the museum director, each is headed by a professional member of the staff. Thus, in addition to the director himself, the professional staff has five members with the rank of curator. To qualify for the position of curator an individual must have a university degree. With the exception of one person, the curators at the Tøjhusmuseet hold either a master's degree or that of doctor of philosophy. This group is augmented by two assistants who are retired members of the Danish Army—one at the rank of colonel and the other a captain. The director himself is a retired lieutenant colonel of the Danish Army.

Forty-two individuals are normally employed in the museum. In addition to the director and the seven members of the staff mentioned above, there are eight technicians within the department of conservation. Each of these is a specialist and the group includes a painter, carpenter, smith, gunmaker, auto mechanic, and specialist in artillery. The administrative and clerical staff has five people—one is a bookkeeper, one a secretary, and three have general duties in the clerical field. The janitorial staff is restricted to four charwomen, but they are assisted by nine guards who also are required to do some cleaning within the museum. However, once a year the museum is closed for a complete cleaning by a private organization.

The museum's operating funds are budgeted for by the Ministry of Defense, with slightly less than 60 percent of the annual outlay designated for salaries. This percentage, which is somewhat below the average for similar institutions, results in part from the addition of two items which do not normally appear in the budgets of European military museums. A certain amount (about 15 percent of the total) is appropriated each year directly to the museum to pay for routine building repairs. In other countries such an item is normally incorporated in the budget of the government agency which is responsible for the upkeep of public buildings. On the other hand, utilities costs are not included in the Tøjhusmuseet budget but are absorbed by the government without direct charge to the museum. Presumably, the cost of any major reconstruction or large-scale renovation of Tøjhusmuseet buildings would not be handled directly by the museum. The Tøjhusmuseet budget also reflects a sum, amounting to about 4 percent of the total, which is allocated for the payment of taxes. The appearance of such an item in a museum budget is most exceptional.

Additional support for the Tøjhusmuseet comes from a group of private citizens known as the "Friends of the Museum." Each year this organization automatically contributes a particular sum for new acquisitions. In addition, however, the organization has considerable capital gained from the contributions of its members, and the museum director is able to draw on these funds to buy additional materials. For example, he may discover a valuable collection that can be purchased for a figure considerably above the yearly allotment. For such a purchase he is entitled to use the capital funds of the organization. However, the organization does have one stipulation—he may draw on its capital only down to the last 25,000 kroner or $3,625. The director states the Friends of the Museum provide an excellent source of support in addition to financial. The group includes many prominent citizens and is headed by a member of the Royal Family. The director expresses great admiration and confidence in the Friends of the

Museum, and he is able to call upon this organization for assistance at any time.

The Tøjhusmuseet enjoys excellent relations with the Danish Army and Air Force. If either service adopts a new uniform, a new insigne, or a new weapon or strikes a medal, the museum automatically receives at least one each free. Because of certain space limitations the museum decides whether it will accept any new aircraft or cannon. Apart from the materials received from these military departments the museum receives no other services from them. The Tøjhusmuseet has no dealings with the navy, for there is a small naval museum in Copenhagen which is supported by the Danish Navy.

The Tøjhusmuseet is fortunate in its location, for the other major museums in Copenhagen are immediately adjacent. All these museums are situated in the heart of the city and so are readily accessible to the public. The Tøjhusmuseet receives many visitors who have a general interest in museums of all kinds. However, the director feels that a large percentage of the visitors to the Tøjhusmuseet come because they are sincerely interested in seeing military objects. Exclusive of the members of the Danish military who come for lectures, approximately 80,000 people visit the Tøjhusmuseet every year.

The present director of the Tøjhusmuseet has been a driving force in stimulating professional cooperation among his European colleagues. Largely through his efforts, the First Congress of Museums of Arms and Military Equipment was held in Copenhagen during May 1957. For the first time in recent years, the curators at military museums came together from most of the countries of Western Europe, the United States, the Soviet Union, Poland, Thailand, and Turkey to discuss topics of common interest. The practical results of the Congress included not only an exchange of useful information and ideas but also a recognition that there is much value in such personal contacts. Before adjourning, the group created the Association of Museums of Arms and Military History and agreed to hold its second meeting in Vienna in 1960. Thus, through its leadership, the Tøjhusmuseet has contributed greatly toward expanding the area of cooperation among the military museums of many lands.

HAERMUSEET
Oslo, Norway

By the beginning of the 14th century the political and geographical center of Norway had shifted sufficiently south and east to make it most convenient for the ruler of the country to establish his permanent residence in Oslo. He also discovered that an invincible fortification was necessary to secure his position, and he chose as the most practical site the rock farthest out on the Akersnes spit of land which juts into Oslo Harbor. On this protected location, Haakon V Magnusson constructed Akershus Castle. The castle has been modified many times during its 650-year history and survives today, through restoration, as the locus around which much of Norway's military history has been written. Outside the walls of the castle are several buildings which have been used over the years to house some of the activities related to the castle's defense. Three of these, of which one was the old arsenal, are now assigned to the Haermuseet (Army Museum).

The Haermuseet was not officially established until 1946, but it has its origins in an Artillery Museum founded in 1860 and a museum displaying uniforms and personal equipment which opened in 1928. After the German invasion in 1940, the two collections were administered by a board of directors (Haermusekomiteen) and other loyal officers. However, in 1941, this group was dismissed and the Nazis placed their own director in control. The uniform and personal equipment collection was evacuated and saved, but the artillery collection located in the Akershus fortress came under German control until 1942. The occupation authorities followed their acquisitive bent and shipped almost the entire collection of metal guns to Germany and the iron guns to a foundry where they were melted down for modern use. German troops stole a considerable portion of the collection of small arms. Many of these items were also sent to Germany by German military authorities. Of the guns sent to Germany, about two-thirds were returned to Oslo after the war through the intervention of the Society of Norwegian Museums. What was left of the Artillery Museum collections was evacuated from Akershus Castle in 1942 and stored in various secret places in the vicinity of Oslo. But misfortune still continued to plague the collection which had been removed from the castle, for a portion of it which had been stored in the old Royal Palace was destroyed by fire.

After the liberation the board of directors and the dismissed officers were restored to control over the collections and resumed their efforts to build a creditable museum. The Haermuseet presently occupies an office building and a storage building, but its exhibit rooms are now being prepared in the old arsenal at the Akershus fortress. The exhibits are expected to be ready for public view during 1961. The Haermuseet also has prepared an artillery park on the drill ground in front of Akershus Castle and adjacent to the museum's temporary offices. It contains a rather sizable group of bronze cannon, howitzers, and mortars and is the one portion of the Haermuseet collection which can now be seen by the public.

The arsenal at the Akershus Castle was constructed around 1700 and is, therefore, of considerable historic interest in its own right. When renovated for museum purposes it should provide an apt setting for the display of arms and armor of a contemporary period. The director of the Haermuseet is planning his exhibits to take full advantage of the arsenal's architecture, particularly for that portion of his collection which dates from the 17th and 18th centuries. The building is sufficiently large to provide the museum with ample room. The entire museum inventory, with the exception of pieces under study by the staff or on display in the artillery park, is located in a separate storage building which is near the arsenal. Much of the material is now crated, but some items are held in a ready reference status. This building appears to be quite adequate for the museum's storage needs and will probably continue to be used for this purpose and to house the workshops. The museum offices are presently located nearby in one end of a former riding school. These quarters are temporary and will be abandoned when the remodeling of the arsenal is completed.

In spite of the losses to its collections sustained during the German occupation, the Haermuseet has a sizable inventory, and cataloguing these items is now one of the principal efforts of the staff. There are several thousand uniforms in stock, including a considerable number of duplicates. The museum plans to discard many of these since they are excess to its needs. The Haermuseet likewise possesses large quantities of medals and insignia which have not yet been completely inventoried. However, the arms have been counted, since they were previously part of the Artillery Museum collections and for the past several years have been the object of much study. The museum has 750 rifles and an additional 700 duplicates, about 2,000 pistols and revolvers, and 600 swords and other edge weapons by type with approximately 1,000 duplicates. The stock of heavy ordnance now consists of 300 guns of all basic types and one armored car. The museum has no military models with the exception of a few in the ordnance category. The remainder of the Haermuseet's inventory consists of 400 flags and colors. Apart from its military collections, the museum has a fine research library which has now grown to about 3,500 volumes.

Much of the Haermuseet program is still in the planning stage, and considerable work yet remains to be done before the museum can open its doors to the public. However, some final space allocations have already been determined, since the renovation of the old arsenal is now being completed. A total of 34,500 square feet is being made available for interior exhibits. Of this amount, about 8,500 square feet will be in a separate building which can accommodate some of the larger objects. The present artillery park containing 6,000 square feet is deemed sufficient to meet the need for exterior exhibit area. Approximately 32,300 square feet has been allotted for storage and is presently in use. The library is now in the same building as the temporary offices and will probably be relocated in the permanent museum structure. The amount of space to be assigned to the remaining museum facilities has not yet been fully decided.

The development of the Haermuseet as a worthy institution for commemorating the achievements of the Norwegian Army has been a governmental project from its inception. When the museum was finally instituted in 1946 through the merging of two military collections, it was placed within the administrative structure of the Ministry of Defense. It actually functions as a semiautonomous agency, but its access to the Minister of Defense is through the commander-in-chief of the Norwegian Army. However, this requirement amounts to little more than the observance of channels for procedural and routine matters. The army also can exert some influence on museum activities if it chooses through its representatives on the advisory committee to the Haermuseet and through the director who is a professional army officer.

The Haermuseet has always had an advisory committee to assist it in matters of major policy and in obtaining needed support from the Minister of Defense for its activities. The group presently consists of five men. Its chairman is the state director of museums who serves as an adviser to all the nation's principal museums, civilian or service-related. Civilian interest in the Haermuseet is given special representation through service on the committee by a prominent civilian museum director. For a number of years this position has been held by Prof. Bjørn Hougen, the highly esteemed director of the Oldsaksamling (Historical Museum) and chairman of the department of archaeology at the University of Oslo. There are three army officers on the committee, and hence the military is in the majority. However, the record of the committee indicates no essential conflict of interest between its civilian and military members. One officer is the chief of the Army Ordnance Corps and the other is the chief of the section which handles uniforms and personal equipment. The choice of these officers seems to be a logical one, since their professional interests are geared to the two major categories of materials comprising the bulk of the museum's inventory. The fifth member of the committee is the director of the Haermuseet, a colonel in the Norwegian Army with a special capacity for museum management.

In the earlier days of the museum the advisory committee was probably more active than it is at present. Museum functions are now more routine and future plans are reasonably well defined, and so the Haermuseet carries on pretty much through its own momentum. The director and the committee are able to work together as a team since the director also serves as a member of the committee. The committee acts upon suggestions offered by the director, especially on matters requiring consultation with the Ministry of Defense or the army. It assists the director in preparing the museum budget and presenting a strong case for increased funds or in requesting other services. The committee is permitted to deal directly with the Minister of Defense. This has proved to be of great value to the museum. Because the committee includes at least two professional museologists, it is competent to offer the director sound advice on technical museum matters. Much of this advice is obtained informally, largely owing to the close friendship among the men involved. This committee, like other museum boards, is empowered to pass on additional policy matters, such as the sale of museum objects, but it delegates much of this to the director. The group has no regularly scheduled meetings, but it gathers at least twice a year in formal sessions. Normally, one meeting is held at the time the annual budget must be considered.

The museum's final administrative organization has not yet been completed. Three departments have already been established, however, and two more are contemplated as the museum is able to expand its activities and trained people can be found to head them. The functions of the three existing departments are defined by the type of military objects with which they deal, and include a department for sidearms, one for firearms, and one for uniforms and colors. The director hopes to create a department of artillery and one which treats army organization. What functions would be included in the latter department have not yet been fully determined. Completing the present organization is the library and a secretariat which are under the immediate supervision of the director.

The museum has only had three directors since it was created. Each of these has been a colonel or lieutenant colonel in the Norwegian Army. All have been military historians and students of weapons and keenly interested in museum affairs. The director's term of office is for an indefinite period; the present director's two predecessors served until their deaths. The director can, of course, request reassignment if he desires.

In addition to the director the Haermuseet presently has only a few full-time employees. Each of the three department heads is ranked as a museum curator and each has been academically trained for his position. Each of the curators has a personal assistant who is a technician or conservator of materials within his particular field of specialization. Two secretaries complete the present complement of regular museum employees. Apart from their routine clerical duties, one looks after the library and catalogue, and the other keeps the files and financial records. The

Haermuseet is also able to call upon some outside assistance on a part-time basis. For example, the librarian of Parliament has served the museum as a consultant on library matters. Since the Haermuseet does not now have a person on its staff trained in library science, such aid is invaluable in keeping the museum's research materials in a condition for effective use. A number of students are also employed part time to work on the catalogues. Thus, the curators are spared some of the more tedious and routine work on the files. The museum does employ one custodian and on occasions has the services of a draughtsman. To what extent the Haermuseet staff will be expanded when the exhibition building is completed is not known. However, an increase is fully anticipated to meet the needs of the rather extensive museum program which has been planned.

The museum already performs a number of services to the public, even though the only thing it is able to exhibit is a portion of its ordnance collection in the artillery park. For example, the curators on the staff make every effort to give complete answers to the approximately one hundred technical questions directed to them each year. There is also a rather well-known association of arms collectors in Norway which receives many questions from the public. Sometimes the association does not have the required information and calls on the museum for assistance. The Haermuseet gladly cooperates with the association in this matter and provides the needed data when possible. The staff identifies a number of specimens which are brought to the museum, but the bulk of the inquiries relate to the conservation and restoration of objects. Because the curators are recognized authorities in their respective fields, they are occasionally asked to give lectures. Most of these are presented to the Ordnance Corps of the Army or to the association of arms collectors. The Haermuseet policy on lending materials is similar to that of other European military museums. This courtesy is not extended to individuals, but duplicates will be lent to other museums or to organizations holding special exhibitions. Some increase in the amount of service the museum will provide the public may be anticipated with additions to its staff and a greater public awareness of its program.

Since the war the curators at the Haermuseet have concentrated on getting their collections organized and in shape for suitable exhibition. Some objects were damaged or allowed to deteriorate during the Nazi occupation, and so the task of conservation has proved to be a rather large undertaking. The staff also discovered that documentation was lacking for many items in the museum inventory. An extensive program of research in the history of arms resulted, and the staff has made some significant contributions in this field over the past decade. Each year the museum includes in its "Annual" another chapter in the series entitled "The Small Arms of the Norwegian Army." Apparently these studies also serve another useful purpose, for the 1951-1952 yearbook notes that "as long as the Museum has no exhibitions, this is the only way of contacting a broader public."

Before the Haermuseet began its research program, very little had been written about Norwegian military arms. Most of what had been written was either inaccurate or too technical to be of much help to museums and collectors wishing to identify their arms. The museum designed its research program both to provide accurate data and to assist those who attempt to identify weapons. The published material thus contains descriptions of all Norwegian regulation small arms together with a short history of arms development covering the same period. In order to make the material understandable for those who are not specialists on arms, strictly technical descriptions have been avoided and considerable use is made of pictures and sketches. Detailed information, however, will be provided later in a series of monographs dealing with single arms or groups of arms. Some of these are already in preparation.

The director of the Haermuseet plans to expand the museum research program considerably after some of the more pressing problems confronting the staff have been dealt with. He feels that a military museum has a vital educational role and that the preparation of studies on weapons, heraldry, and other aspects of military history are most essential. He hopes, therefore, to make the Haermuseet a leading research institution among military museums.

From its beginning the Haermuseet has enjoyed strong official support. This has been reflected in a constantly increasing budget. In the past ten years funds available to the museum have more than doubled. Some of the credit for this steady increase must go to the advisory committee, for its members, particularly the civilians, intercede directly with the Ministry of Defense in behalf of the museum. Special funds will be made available to the Haermuseet when its exhibits open in 1961, and further increases are anticipated after that. The museum is fully financed by public money, and the annual allotment is programmed as part of the army budget. Funds for building construction, repairs, and maintenance are provided for separately and hence do not appear as part of the museum's regular budget.

The Haermuseet is more closely identified with its parent service than are many similar institutions in other countries. The Norwegian Army's representative on the Museum Advisory Committee and the choice of an army officer as director give the army some influence in policy matters. But the army also demonstrates its interest by contributing a number of specimens and performing some services for the museum. The army gives the Haermuseet at least one of every new weapon or uniform it adopts together with whatever materials the museum wants from its old stock. The museum also receives assistance in its conservation work, some funds to be applied for its electricity and help with its transportation needs.

The Haermuseet now works within an aura of expectation. Its immediate goal is to prepare worthy exhibits for public view. Its basic desire is to provide a complete picture of the Norwegian Army and to bring credit to itself as a responsible research institution. In the future, it can also be reasonably sure of much public attention, for those who come to visit Akershus Castle, one of Oslo's most celebrated tourist attractions, must pass by its door.

ARMÉMUSEUM
Stockholm, Sweden

Although the Swedish people have not been involved in war for almost a century and a half, their earlier history was crowded with international conflicts, and they developed a number of lasting military traditions. They have likewise made significant contributions to technical developments in firearms, swords, heavy ordnance, and field engineering. In years past they coped with the problems of land defense and constructed elaborate fortifications which still remain of great historical interest. These military traditions and achievements have been given worthy commemoration in the Armémuseum (Royal Army Museum) in Stockholm.

The Armémuseum is one of the oldest military museums in Europe, for it first opened its doors to the public in 1879. Its location was the principal building of the Main Artillery Depot, a structure which it still occupies. The museum's origins actually go back many years earlier, for during the 17th century a collection of ordnance trophies was first displayed in the Artillerigarden (the Ordnance Court) at the same site. This early collection of cannon gradually took the form of a museum when it was augmented by the collections of edge weapons and firearms gathered during the reign of Charles XII and the "Cabinet of Models for the Science of Gunnery" built around 1750. A little over a century later, in 1877, all the military collections belonging to the various institutions of the army were drawn together into one museum, the Armémuseum. Two years afterward, the displays were first made available to public view. The museum has in no sense been static and has grown considerably during its 80-year history. For example, it has added a large and representative collection of army trophies and ordnance objects from several countries.

The Artillery Court in which the Armémuseum is located has served as the central depot for the Swedish Artillery from 1641 until the present. The building housing the military collections which stands along one side of this large court, was erected during the years 1766-1769. When first constructed it contained two floors, but this space proved too small to meet the needs of the museum and a third floor was added in 1883. The enlarged building was made even more impressive with the addition of a central dome similar in appearance to the one which tops Napoleon's Tomb at the Hôtel des Invalides in Paris. From 1932 to 1943 the majority of the exhibition halls were modernized, and the museum attained its present state of attractiveness, although it still suffers from a number of limitations imposed by age and a lack of sufficient space. Its workshops are quite small. It has been able to allot only a minimum of space for service facilities. The museum has no auditorium, and it is unable to provide such public conveniences as a refreshment room or sales room.

Access to the Army Museum is through the Artillery Court, part of which is used for the museum's exterior exhibit. Immediately in front of the building and to either side of the entrance is a long row of trophy cannon mounted on low runners which keep them a few inches off the ground. Just in front of the cannon are several small carriage-mounted field pieces. The cannon date from the 17th and 18th centuries and contain the elaborate decorated ornaments in relief which characterized the ordnance of that age.

The ground floor, which contains the artillery and engineering halls of the museum, architecturally reveals the building's long use as an ordnance depot. Both halls are contained in a single room which runs the full length of the building. The room itself is characterized by four rows of large square pillars which support an impressive vaulted ceiling. The front windows of the building admit the only natural light, but at every corner of each pillar is an electric light arranged so as to provide indirect lighting. Thus, the room is assured of a fair amount of brightness. The cannon, field pieces, and other objects are lined against the walls and displayed chronologically. The wide central aisle is left free of exhibits with the exception of a cannon and a horse-drawn artillery piece. The monotony of the central aisle of pillars is broken occasionally by a small artillery tube, mortar, or an array of cannon balls placed beside them. A large number of pieces are on display closely together, but they can easily be examined on all sides. The room looks like an arsenal, as undoubtedly it once was, and produces a proper atmosphere for a display of ordnance.

The second story of the museum, like the ground floor, is a single room which is almost as long as the building. This floor also contains the museum offices. The ceiling of the room is flat but is characterized by large cross beams which provide an ideal place from which to suspend colors and standards. Two central rows of wide arches parallel the full length of the room and have been used most effectively for display purposes. Several of the archways have been closed off and glassed in as exhibit cases. Most of these special cases contain groups of uniformed figures, although other objects are also on display. The wall space between the arches is likewise used effectively. In some instances pictures are hung from the wall, or the space contains a small rack of swords, rifles, or halberds. Some of the open archways also have a small case in the center which contains one or several items of unusual merit. The walls of the room are lined with cases, gun racks, pictures, and other objects and are grouped in special exhibit units. One of the major exhibits is the Bernadotte Hall, but most of the displays deal with personal arms and uniforms in use by the Swedish Army since 1818. Another special feature is "Today's Corner," an area which also is used for temporary exhibitions. Like the ground floor, the second is well lighted, and the floor space is almost completely free of exhibits. The viewer easily acquires an impression of spaciousness as he gazes its full length.

The third floor was obviously constructed with a view to its use as a museum. Its ceiling is considerably higher than the lower two floors, and it has been divided into a number of separate, but interconnected, rooms. It is, thus, much more adaptable for the chronological presentation of Swedish military history, and individual halls have been prepared which date from the early 16th century to the Napoleonic period. The Armémuseum has constructed its historical exhibits around the exploits of Sweden's great soldier kings. The chronology begins with halls displaying objects from the early and late Vasa periods which correspond to the mid-16th and early 17th centuries, or the reigns of Gustavus I and Gustavus Adolphus. Separate halls are also designated for Charles X, Charles XI, and Charles XII. In addition, the third floor has several topical halls. One provides a display of significant trophies, another exhibits a fine collection of personal arms and uniforms used by Swedish troops up to 1809, and one features some 30 fortress models, most of which date from the late 17th century. Individual halls also contain a number of flags and colors in use during contemporary periods. Full advantage of the high ceiling is taken in displaying these colors, for they are suspended from poles attached to the upper molding and hang flat without any folds in the material. This method of display permits the visitor to walk beneath the flag and see its complete design from either side. The preservation work on all the flags is excellent. The same exhibit plan is used for this series of halls as for those on the lower two floors. The cases, racks of small arms and edge weapons, and other objects are all placed against the walls, leaving the entire center portion of the rooms completely free of any artifacts. For this reason the halls strongly reasemble a fine-arts display.

The exhibits of the Armémuseum attain a very high caliber. In its program of modernization the museum staff has apparently felt it should make a visit to the museum an entertaining venture for the public. The displays are impressive and have been accomplished in excellent taste. The objects have been selected with care to insure that only the finest pieces are on view, and these reflect use of the best preservation techniques. Many individual objects and collections might be singled out for special comment, but a few generalizations will suffice.

With the possible exception of its heavy ordnance, the museum displays only representative artifacts. The portion of the inventory in storage is not known, but the amount of space in the building reserved for this purpose is quite negligible. Thus, virtually all the cannon and other artillery pieces are exhibited on the ground floor or in the Artillery Park. The exhibit provides very comprehensive coverage for the evolution of artillery together with representative samples of cannon cast in Sweden and illustrations of innovations which Swedish designers have introduced.

The museum displays are not encumbered with row upon row of weapons and uniforms, although there is an ample supply of all types. These items are used in two ways. In the historical exhibits, they are utilized in combination with other materials to chronicle the military events of a particular period. In some of the topical halls they provide the subject material. Here weapons are displayed by type, with examples of variations together with those which highlight significant changes occurring during their evolution. The various displays of uniforms within the Armémuseum are particularly superb. The garments are in an excellent state of repair, and most have been placed on lifelike appearing manikins. Because these are of such fine quality and have been used so effectively in well-designed exhibits, the museum's uniform displays are probably the best in Europe.

In developing their displays the Armémuseum staff has carefully avoided the pitfall of monotony. Each floor has its distinctive features, and the architecture of the building has been used to the best possible advantage. As a result, a viewer is inclined to admire the ingenuity of the museum designers as well as the fine quality

Armémuseum, Stockholm. Liberty Period Hall (1719–1772).

Armémuseum, Stockholm. Part of the Bernadotte Hall.

Armémuseum, Stockholm, Infantry uniforms, 1820's-1830's.

Musée d'Art et d'Histoire, Geneva. Hall exhibiting armor and scaling ladders.

of their exhibits. Throughout there is a good mixture of topical halls and those designed to portray a specific period of military history. The record of the Swedish Army is well commemorated throughout the museum, not just the exploits of warrior kings or great commanders. Kings are memorialized through the halls that bear their names, but these halls are not restricted to a display of personal relics. Rather they portray the achievements of the King's army. Of course, personal mementos are there, and these prompt added interest in individual exhibits.

There is an excellent use of space throughout the museum. No crowding is evident, with the possible exception of the orderly rows of heavy ordnance on the ground floor. Even here, the space is used effectively; no object is obscured from thorough examination. The museum does have one rather serious space limitation. One area on the second floor is reserved to display items in current use by the Swedish Army. It also doubles as the room assigned to handle temporary exhibitions which are held on an average of two or three times a year. As a result, permanent exhibits have to be moved periodically, especially when a fairly elaborate special display requires considerable space. The Armémuseum director cautions that any new museum house should provide a special space for temporary exhibitions, for he believes that these are required nowadays in military museums because of their increasing popularity.

Except for special exhibitions the displays seldom change. New materials are incorporated as soon as they have been properly conditioned and the necessary research upon them has been completed. Descriptive material for the objects is held to a minimum. Thus the labels are easy for a viewer to follow because they are kept simple. The visitor is also given some instruction on his tour of the museum through the catalogue, but for the most part he need do little more than observe the exhibits carefully to acquire a fair knowledge of the essential facts about Swedish military history.

One objective of the Armémuseum has particular merit, i.e., a desire to be fully contemporary. The staff apparently feels that the visitor not only is interested in past history but also desires to know something about the Swedish Army of today. Such knowledge is conveyed in "Today's Corner." In this room the latest equipment in use by the army and what will be standard for some time to come is shown. Since the museum must cope with the problem of security restrictions in this exhibit, a special army committee inspects it to see if it is in order before the public is permitted to view it.

Most of the military museums in Europe give little attention to the contributions of their nation's military establishment to the field of medicine. Some do have a few objects, such as ambulances, field medical kits, and surgical instruments, but seldom are these integrated into a special exhibit. The Armémuseum took the opportunity afforded by the 150th anniversary of the first medical unit in the Swedish Army to prepare one of the largest special exhibits it has ever attempted. Its displays portray significant events in the history of Swedish military medicine and demonstrate thereby that military museums can effectively present the valuable contributions the armed services make to this vital field. The exhibit opened in October 1958.

The Armémuseum's concern with public interest is evident not only in the attractiveness of its exhibits but also in other aspects of its program. Its chief objective is sustained, not sporadic, public interest. Hence, its special exhibitions follow one another rather closely. There is an occasional motion picture shown in one of the halls, but the museum is limited in what it can do with this medium of instruction or with public lectures because it has no auditorium. For a military museum it reaches an ultimate in public entertainment by holding a concert the second and fourth Sundays of each month. Usually a band provides this entertainment and features martial music in its program. This attraction has increased somewhat the number of visitors. Guided tours are also available for the public, and the staff provides the usual service in identifying objects or offering advice to amateur collectors. Other inquiries received from the general public are handled rather informally. Complex questions requiring some detailed research number only 40 to 50 a year. Lectures by the staff are normally given only to members of the army, army cadets, students at the Artillery School, and other selected units who come to the museum for lectures on special subjects and a tour. There are usually 10 to 15 such tours a year.

The museum maintains a fairly vital academic atmosphere and publishes a considerable number of studies prepared by its staff. The principal publication is a yearbook which contains articles written by staff members and other students of weapons and military history. An account of the annual activities at the museum is also included. The emphasis given to research at the Armémuseum is largely inspired by the director himself, who is an outstanding military historian and the author of several highly regarded books in the field. The bulk of the research done by the staff is, of course, routine, and performed in conjunction with their work upon the collections.

The Armémuseum has an inventory containing approximately 30,000 numbered items, which is slightly in excess of that possessed by the Heeresgeschichtliches Museum in Vienna. The bulk of its arms, uniforms, and insignia collections are Swedish in origin, but they also include a sizable number from other countries, some of which were obtained as trophies of war. The weapons collection includes 4,500 small arms, almost 400 of which come from outside Sweden; approximately 1,300 swords, sabers, and other edge weapons; 400 cannon and mobile artillery; and 8 armored cars. The museum has over 1,700 uniforms, one-fourth of which come from other countries, and 1,100 insignia and medals. The collection of approximately 430 colors is almost equally divided between those carried by Swedish units and those captured from the enemy. There are also more than 100 ordnance models, over 30 models of fortifications, and several engineering models. The Armémuseum has a large number of pictures and other documents which are incorporated into its displays. It likewise is well equipped for its research activities with a reference library presently containing about 10,000 volumes. The library is primarily for the use of the staff and specialists, but as far as possible it is made available to the public for study purposes.

By European building standards, the Armémuseum is medium sized. It currently has 45,700 square feet allocated for interior exhibit, an amount similar to that of the Imperial War Museum in London, the Musée de la Marine in Paris and the Museo del Ejercito in Madrid. Its exterior exhibit space located immediately in front of the museum has an area of 14,100 square feet. Its maintenance and workshops are fairly small, occupying only 1,100 square feet, and its offices are housed in a space of 1,300 square feet. The library is also somewhat limited in space, and is in a single room containing an area of slightly more than 500 square feet. Separate facilities are provided for study, but this is restricted to one small room.

Since the Armémuseum's founding it has been a public enterprise and has been an agency within the Ministry of Defense. It is the only military museum in Sweden which is under the supervision of a military department. In respect to the control of its military museums, Sweden runs somewhat counter to the prevailing European pattern. Usually, such museums are administratively related to the appropriate service ministry of the nation concerned, seldom to any other department of government. Thus, the Armémuseum is the exception to the rule employed by the Swedish Government in fixing the jurisdiction of its service-related museums. In exercising its control over museum operations, the Minister of Defense has maintained a similar degree of aloofness as that found in most European nations. Normally, it has restricted itself to reviewing the budget and policy decisions of the museum's governing body, although occasionally the Minister of Defense will appoint a new secretary who takes considerable personal interest in Armémuseum affairs. When this occurs the result has been to disturb somewhat the normally smooth course of museum-ministry relations via the avenue of bureaucratic disruption.

To some degree the director of the Armémuseum is spared the necessity of assuming full responsibility for adjusting to any change in relations between the museum and its parent agency. He is assisted in all matters of major policy by a board of directors, although he likewise is one of its members. The board has five members, four of whom serve in an ex-officio capacity. Three of these are appointed from the army, and membership is an auxiliary duty devolving from their regular assignment. The three army appointments are the chief of staff of the Swedish Army, the chief of Swedish Artillery, and the chief of the unit in the army which handles uniforms and equipment. The designation of these officers is somewhat logical in view of the great emphasis which has always been placed upon the museum's collections of ordnance, uniforms, and personal equipment. The fourth ex-officio member of the board is the director of the Armémuseum who serves also as its secretary. The fifth member is an eminent civilian appointed for an indefinite term by an official known as the chief of antiquities. The present civilian representative is a distinguished scholar. As is somewhat evident from their method of selection, the members, other than the director, are not experts in museum management. Hence, they depend fully upon the director to run the museum and content themselves with a strictly advisory role. They assist him on policy matters in a very general way, pass on the proposed budget, help the director in obtaining objects he desires, and serve as a source of valuable support whenever the need arises. The directors meet only twice a year.

At present the substantive functions of the Armémuseum are performed within the administrative framework of three departments. The division of work is identical with the major categories of specimens within the museum collections. Hence, there is a department for uniforms, one for colors, and a single department for all weapons. Each of these is headed by a curator who is a specialist in a particular category. There is no formal service department, but the technicians are under the immediate supervision of the director, as are administrative and clerical personnel. Thus, the Armémuseum is organized with a great deal of administrative simplicity and as a result seems to be managed most efficiently.

Even though the Armémuseum has a sizable collection, a fairly large exhibit area, and a most ambitious program, it has comparatively few employees. Usually the staff has only 10 full-time members. The professional group consists of the director and three curators, each of whom heads a department. One of the curators also has an assistant. The chief technician, an armorer, is assisted by two carpenters and a painter. One other technician works on the restoration of flags and colors, a field of activity in which the Armémuseum excels. The director has a secretary who also serves as the museum's cashier. There is also an assistant in the office who works half days. The museum has one janitor who is assisted by a part-time employee, but these are not considered part of the regular staff. The Armémuseum has no maintenance personnel, for work of this type is performed under contract.

The major source of the museum's financial support is from public funds. Its operating or routine expenses are budgeted for through the Ministry of Defense and appear as appropriate entries in the over-all budget of the ministry. Utilities and maintenance costs are paid through a special budget, and funds to cover these items are not appropriated to the museum. Thus, the financial affairs for the Armémuseum are handled according to the normal European pattern. In the case of the routine budget, about 75 percent of the money goes for salaries, 20 percent for general upkeep, supplies, and travel, and 5 percent for conservation and acquisitions.

The museum has three additional sources of income, but the total amount realized is not too large. The Association of Friends of the Museum provides some money for the purchase of new objects. The amount given to the museum varies each year normally contingent upon the status of the association's treasury. The museum charges a nominal fee for admission, and the income is placed in a special fund which may be used for the purchase of supplies. The amount raised through this method is not very great, since all members of the services in uniform are admitted free, and the museum is free to everyone on Sundays and holidays. A final amount of money comes through the sale of publications. Again the sum is small, but any profits realized are retained by the museum to be used as seen fit.

Since its founding the Armémuseum has maintained a rather close relationship to the Swedish Army. From the standpoint of official interest, the army is well represented on the board of directors and apparently offers no opposition to civilian direction and staffing of the museum. The army also expresses its interest in other ways. With the exception of artillery, it donates one of every new item it adopts for use. It likewise gives any surplus items or objects from its obsolete material. It augments the staff of the museum with three enlisted men who are sent to work at tasks the director assigns to them.

The Armémuseum gives assistance to other army museums or military collections in the country. It lends objects or portions of its collections to these local institutions, restores colors for them, and performs any other conservation work required. It likewise offers its services as a consultant, and its director has the added responsibility of inspecting them.

The Armémuseum enjoys the advantage of a good location, for it is in the heart of Stockholm, only three or four blocks from the intersection of two of the city's main streets. By comparison with other European military museums located in cities of similar size, its number of visitors is not large. The present average is about 25,000 a year. It is somewhat difficult to discover possible causes for the relatively small attendance at this exceedingly fine museum in light of its efforts to stimulate public patronage. Perhaps part of the reason may be found in the competition offered by other outstanding museums in the city, the specialized nature of its collections, or lack of interest in a military museum by the citizens of a nation which has managed to avoid the holocaust of war for over 150 years.

STATENS SJOHISTORISKA MUSEUM
Stockholm, Sweden

As was emphasized in discussing the Musée de la Marine of Paris the major navy museums of Europe do not restrict their attention merely to military matters. Rather, they tend to treat the full sweep of maritime activity. Actually these museums provide about equal exhibition space to the navy with its war and peace-time missions and to the merchant fleet and related mercantile activities. The visitor is reminded not only that exploits on the sea have been performed by the man-of-war, but also that the merchant vessel has established an illustrious, though often less spectacular, record of prowess as it traversed the sea lanes of commerce. In each instance, the museum has approached its task from the viewpoint that the ocean has been an area of great national activity and that all aspects should be considered in one institution. The result has been most satisfactory, for where this treatment has been attempted a museum of high quality has resulted. The National Maritime Museum in Greenwich and the Musée de la Marine in Paris are outstanding examples of such museums. To them must be added the Statens Sjohistoriska Museum in Stockholm, which was established to provide opportunities for study of the history and scope of the Royal Swedish Navy and the Swedish Mercantile Marine.

The Statens Sjohistoriska Museum (National Maritime Museum) is a comparatively new institution, having been opened as recently as 1938. Planning for this museum began a number of years earlier, and construction of a building was finally started in 1933. The building was completed two years later, and some three years more were required to ready the exhibits. The collections which are now the property of the museum were assembled from several sources, and many of the objects are of considerable antiquity. For example, the exhibits in the Royal Navy section were selected from those existing national collections relating to the history of naval warfare which passed to the Sjohistoriska Museum. Some were presented by private collectors, some had been in the Naval Yard Museums at Karlskrona and Stockholm, and the other pieces had been in the possession of local naval authorities and deposited in training or administrative establishments. The exhibits in the Swedish Mercantile Section came originally from the Swedish Shipping Museum Association, which had been founded in 1913.

The building housing the collections of the Sjohistoriska Museum has a distinction quite rare among military museums in Europe; it was actually designed and constructed for museum purposes. Because it was built so recently, its exhibit halls are modern in design and respond exceedingly well to use of the latest museum techniques. The exterior of the building gives the appearance of plain and simple dignity. The ample grounds around the museum are beautifully landscaped, with the front of the building facing a broad grassy expanse. The two-story building is entered through a circular central foyer which juts out in front of the wings at either side. Just back of the entrance hall is a spacious rectangular room which serves as a multipurpose hall and reaches a number of feet beyond the back walls of the museum wings. The two symmetrical wings curve forward slightly until at the end they reach a point parallel to the front entrance. This arrangement strongly resembles the curved halls of the Palais de Chaillot in Paris, although in the case of the Palais the curve is much more pronounced. Since the two buildings are of a contemporary period, it is possible the architect of the Sjohistoriska

Museum may have been influenced to some extent by the design of the Palais. On the other hand, he may have chosen to duplicate the arm of an anchor, feeling that such a shape would appropriately symbolize marine activity.

The museum houses the exhibits of its two principal subject areas on separate floors. The collections in the Royal Navy section are displayed on the ground floor, those in the Mercantile Marine section are located on the upper floor, and several additional exhibit rooms have recently been constructed in the basement. The attractive entrance hall or rotunda serves to orient the visitor to the exhibit plan and contains the staircase to the upper floor. The catalogue, which is available at the entrance, contains diagrams of the exhibit floors and suggests the direction the viewer should take in his tour.

The Royal Navy section, which is the military division of the Sjohistoriska Museum, has a chronological presentation. Individual halls trace the evolution of the Swedish Navy from 1521 (the beginning of the Vasa period) until the present, and they generally feature objects belonging to precisely defined periods in Swedish history. Four of the halls, however, are topical in nature, but these appear at the appropriate point in the exhibition's historical context. The periodical division of the historical rooms is similar to that in the Armémuseum, and both museums begin their story of Swedish military power with the ascension of Gustavus Vasa to the throne. His reign marked the emergence of Sweden to a position of national prominence. Other rooms which center their displays around particular kings are the Caroline room (1680-1772), the Gustavian room (1772-1815), and the Oscar room (1880–1907).

The remaining halls in the navy section are concerned with key developments or transitional periods of the navy. For example, the evolution of the Inshore Navy (1756-1815) was a significant epic and began with the assignment of oar-propelled flotillas to the War Office. By 1788 it included a great number of small gunboats and attained its zenith in 1790 when it defeated the Russian Fleet in the celebrated battle at Svensksund. The achievements of the Inshore Navy are appropriately commemorated in a special room.

The period 1815-1880 is covered in two adjacent topical halls named the mast room and the battery room. These 65 years were most significant ones for the navies of all countries, for they marked the change from sail to steam, from wood to iron, and from the round shot of the smooth-bore gun to the shells of the rifled barrel. The mast room exhibits models of sailing ships appropriate to the early part of the period. It provides particulars about the disposition of the fleet and details about the principal expeditions conducted during the period. The battery room contains the reconstruction of an original battery from a 19th-century frigate together with models of steam ships. It also includes showcases depicting naval logistics, care of the sick, small arms, and the development of certain weapons.

The Sverige room (1907-1939) demonstrates the evolution of Swedish battleships through construction of the larger, faster, and heavier-armed Sverige-class vessels. In addition, the hall contains models of aircraft the navy employed before the Fleet Air Arm was incorporated into the air force in 1926.

The period since 1939 is treated in two rooms. One is concerned principally with the mobilization of the Swedish Navy in World War II and contains examples (mostly models) of ordnance used at that time. The second room affords a few glimpses of the Royal Navy today, including a display which treats the theory of guided missiles.

The largest and most impressive room which is located on the ground floor at the center of the building is designed to serve as a suitable memorial to the valor of the Swedish fleet in battle. It is the only hall in the museum which is frankly commemorative and is called appropriately the hall of remembrance. Because Sweden has fought no wars since 1815, most of the room's exhibits date back to the Gustavian period. It is two stories in height which permits the mounting of large historical figureheads around the walls. Jutting through the back wall of the room is the actual stern of Gustav III's schooner *Amphion*, the vessel from which he directed the campaign against the Russians in the war of 1788-1790. The interior of the cabin can be seen by entering the small room to the rear of the hall of remembrance. At the exit of the hall are two bound volumes containing the names of all men of the Royal Navy and Mercantile Marine who lost their lives at sea during the second World War. The tribute to these men is all the more significant inasmuch as Sweden maintained its neutrality throughout the entire conflict. The hall is used for a number of purposes because of its size. For example, it serves as the museum's auditorium, and it has been used for receptions and various other special affairs.

Although the exhibits in the Mercantile Marine section on the upper floor are not of primary concern to the student of military museums, they do demonstrate useful examples of excellent display techniques and provide an interesting treatment of topical subject matter. The chronological pattern of presentation used in the Royal Navy section is followed to some extent in this series of halls. At the beginning of these exhibits some effort is made to illustrate Viking and medieval commercial activity, but the bulk of the displays deals with Swedish commercial enterprise from the early 19th century to the present. The evolution of Swedish merchant ships over the years is evident as the visitor moves from hall to hall. Excellent ship models are featured throughout. The exhibits include models of early and modern shipyards and segments of masts and full-scale figureheads, and one contains models of ships used by the principal Swedish shipping lines of today. Special displays present the development of ships' machinery, navigation instruments, compasses, spy-glasses and telescopes, seamen's training and organizations, and examples of hobbies and souvenirs of Swedish sailors. In the attention it gives to the life of the average seaman, the Sjohistoriska Museum has attempted something quite unique for this type of European museum.

The museum expanded its coverage of mercantile activities by developing a series of exhibits around the subject of ship construction. The presentation is an ambitious and attractive one, for it features full-scale cabins and other facilities aboard ships of various types. The viewer may walk into the cabins and gain the impression he is actually aboard ship.

Because the museum building was constructed for its present purpose, its interior has been designed to provide a series of interconnected galleries. The visitor may thus follow a precise plan during his tour. At no point must he retrace his steps to get from one part of the museum to another. This has been accomplished by dividing the wings on the ground and upper floors into two parallel sets of galleries. A visitor enters the first hall on each floor and by touring each room in order on that level returns to his starting point. The rooms throughout the entire building were planned as to ultimate use in advance of construction with the idea that, once installed, the exhibits would be fairly permanent. Thus, the exhibit halls vary considerably in size and shape.

The exhibit designers reveal a keen esthetic sensitivity. Throughout they have capitalized upon the attractiveness of naval artifacts, particularly ship models, and their displays assume a pronounced fine-arts character. The need for maintaining a historical context is met by providing several diagrams and charts illustrating the strength of the fleet, maps of battles and naval campaigns, pictures of celebrated commanders, and succinct labels.

Most of the cases in individual rooms have been placed against the walls. Occasionally there is a showcase in the center of the room or away from the walls, but the usual pattern is to place such cases adjacent to and at right angles to one of the walls. There is no overcrowded room in the museum, although it is fair to say that only a few visitors at a time can be accommodated in the rooms which represent reconstruction of actual ship facilities. Elsewhere space is not at a premium. Some attention is also given to visitors' comfort, for a number of halls have seats where the viewer may pause to rest.

It is perhaps axiomatic that any naval museum will center a goodly number of its exhibits around ships models. The Sjohistoriska Museum has an exceedingly fine collection and has sought to display them to best advantage. For the most part, the models are in cases, but a number of the larger ones are not. These have been mounted in a simulated ocean setting to give the impression the ships are under sail. To furnish the viewer with some concept of size, small, carved figures of sailors have been placed aboard and are shown in the act of performing a number of ship's duties. Thus, the models make a lasting impression upon the visitor and contribute much to the museum's high standard of quality.

Beyond a display of weapons and small items used on board ship, it is most difficult for a naval museum to exhibit real objects. Obviously anything larger than a barge or small boat cannot be brought inside the museum if other items are to be displayed. The staff at the Sjohistoriska Museum has sought to circumvent this problem somewhat by constructing its series of important ship elements to actual scale. The stern of the *Amphion* at the rear of the hall of remembrance brings the visitor to the actual cabin where Gustavus III held councils of war. The ship's battery and reproductions of various cabins introduce much realism into the displays. These are augmented by the many actual figureheads from famous ships and the bases of masts from old sailing vessels complete with their riggings. The value of using real objects or those constructed to proper scale should never be underestimated as to the vital contribution they make to the attractiveness of the exhibits, the sense of participation in past events they convey, and the valuable educative technique they provide.

Throughout all its displays the Sjohistoriska Museum emphasizes the showing of representative objects, rather than a multitude of specimens. Weapons are exhibited by type and the ship models often illustrate particular types of vessels that have been in the service of either the Royal Navy or the Mercantile Marine. The weapons are attractively mounted in showcases, but in general the visitor is restricted to a view of only one side. By showing only significant items in its collections the museum keeps the interest of the viewer during this whole tour, accords itself considerable flexibility in developing attractive displays, and does not detract from the admirable qualities of its significant objects.

Some changes are made in the exhibits, but these are usually limited to the addition of new objects or the making of corrections where further research reveals errors. The hall which treats the navy of today must, of course, be kept contemporary. An occasional new exhibit is built, but space available within the house for such purposes has now been almost completely used. There is no room specifically assigned to set up temporary exhibits. The only place where this can be done on any scale is in the hall of remembrance.

It is evident that the Sjohistoriska Museum is interested in maintaining strong public support for its activities, but it depends largely upon the high quality of its exhibits to sustain this. The remaining services it offers the public, however, are about normal for museums of comparable size and resources. The amateur arms collector and the student of naval history are able to draw upon the assistance of the staff in several ways. The staff attempts to answer questions of moderate complexity, but this assistance is given informally and on a limited basis. Slightly over 100 of these inquiries are handled each year. Specimens are identified when brought to the museum and some technical advice is furnished the private collector. Authors of books and articles may borrow from the Sjohistoriska Museum's complete collection of Royal Navy photographs, but the museum lends nothing more from its collections to the general public. It should be noted in passing that the museum extends this courtesy on infrequent occasions to other military museums and sometimes will borrow objects to augment its own displays. It likewise will repair ship models for churches and other institutions where such objects are kept on display and will serve in a consultative capacity for other Swedish museums if requested to do so. The museum does not conduct lectures for the general public, but infrequently it does so for the Society of Friends of the Museum. The same applies to the showing of motion pictures. The hall of remembrance is used for these gatherings. Guided tours are not normally available, except for clubs and private groups, and there is a charge for this service. However, in the case of school groups the service is free, and the tour is conducted by one of the guards.

The Sjohistoriska Museum has an excellent library and archives which contain valuable research materials for the staff and other students of Swedish naval history. In earlier days the museum maintained a special research division, but its current administrative structure does not now provide for a separate research unit. Students who come to the museum for study are permitted to use the materials available in the library and archives and may depend upon assistance available through the staffs of both. The research performed by the museum's professional staff is largely limited to that done in connection with their routine work.

The Sjohistoriska Museum is still a comparatively new institution, and much research activity remains centered around the objects in its inventory and those which are added continually. The curators have their fields of personal interest and attempt to do intense specialized study, but the time available for this is yet quite limited. The independent inquiries now undertaken are inclined to be restricted in scope, and thus staff members have produced no major studies. Apart from its catalogue, which is revised periodically, the museum produces no monographs or specialized studies. However, it does publish a biennial volume which provides some outlet for museum research. In addition to a description of significant activities at the museum, this publication contains a number of special articles written by staff members and guest authors. The curators also are men possessing academic status. In this capacity, they maintain an association with Stockholm University.

Exclusive of its archives and library, the Sjohistoriska Museum has a current inventory of approximately 11,600 items. The greatest portion of its objects cannot be classified as exclusively military in nature. These include navigation instruments, telescopes, many other objects used aboard ship, pictures, figureheads, and some flags and pennants and account for about 9,100 items. The remaining objects are almost equally divided between models and full-scale specimens. The museum has a stock of about 200 uniforms and 390 insignia and medals, but only a few of these specimens are on display. The weapons collection includes approximately 250 small arms, 330 swords and other edge weapons, 170 heavy guns and cannon, and 45 mines and torpedoes. There are about 500 ship models in the Sjohistoriska Museum's military collection; this is about half the number possessed by the National Maritime Museum in Greenwich. The museum also has 325 ordnance models, 15 aircraft, and about 300 models of other types of naval equipment.

From the standpoint of interior exhibition space, the Sjohistoriska Museum is one of the smaller military museums in western Europe, for it allocates only 13,900 square feet for this purpose. However, the building has been planned as a compact unit with very little waste space. The coordinated galleries have been designed to provide a great deal of wall area. This in turn permits the efficient use of showcases and eliminates many of the problems other museums encounter in converting inadequate rooms of considerable size to exhibition halls. The stairways and corridors in the entire building contain only 650 square feet. This indicates the high degree of effectiveness to which space is utilized within the museum. Thus, the staff members have been able to place more in less space than have many of their less fortunate colleagues, and still they have kept the galleries free from overcrowding. There is also approximately 17,500 square feet of exhibition space available in the exterior area adjacent to the building. However, this is little used. There are a few mounted cannon, a ship's anchor, and a mast and rigging from a sailing vessel outside. These are attractively placed and do not detract from the fine landscaping around the building. Of the almost 15,000 square feet of storage space available to the museum, about 700 square feet is used to house reference collections under current study, and the remainder is used for dead or permanent storage. The workshop and laboratories are of average size and occupy approximately 1,500 square feet. The library and archives contain 1,500 square feet, the museum offices occupy 1,700 square feet, and the remaining service rooms have about 7,000 square feet of floor space.

Unlike its counterpart museum for the Swedish Army, the Sjohistoriska Museum is not linked to the Defense Department. Rather it is an agency within the Merchant or Commerce Department. The reason for this relationship is somewhat obscure, but it is not at all illogical, since the museum covers the full range of marine activity in its exhibits and greatly emphasizes Swedish maritime commerical enterprise in its displays. In the attention it gives to the management of the Sjohistoriska Museum, the Merchant Department follows the usual pattern of European bureaucratic control. It largely restricts its concern to budget matters with the automatic annual fiscal review, and leaves the more technical aspects of museum operations to the staff and board of directors.

The 8-man board of directors, which develops the major policies of the Sjohistoriska Museum in most general terms, regards itself primarily as an advisory and intermediary body. In accordance with the terms of its assigned powers, the board must review the budget and submit it to the Merchant Department for ultimate approval and inclusion in the annual budget for the whole department. The board, of course, has the option of exercising closer supervision over museum activities, but its record from the very beginning has been essentially one of steady support for the director, who is also a member of the group, and ratification of his decisions. It has always deferred to him on technical and routine matters. It never inspects the museum for purposes of supervision, but the members do visit the exhibits to keep abreast of current activities. Meetings are held four times a year.

Membership on the board has been provided for the representatives of certain public and private interests. The museum's two principal subject areas are represented by members appointed from the Royal Navy and the Merchant Navy. The director is a member both as a matter of convenience and to give the board immediate consultation with a museum expert. The interest of private citizens who form a nucleus of loyal support is represented by a member drawn from the Association of Friends of the Museum. By law, these four individuals must serve on the board, but additional members are added by invitation. At present the board has representatives from the Department of Justice and from the Dockyards which is considered a separate government agency and not within the jurisdiction of the Merchant Navy. All members, with the exception of the director who serves ex-officio, are appointed by the head of the agency represented. The commander-in-chief makes the appointment in the case of the Royal Navy and the Merchant Navy. There are no set qualifications for service on the board. Normally the appointed representatives have a known interest in naval and merchant marine matters, and have also demonstrated some interest in general cultural advance. They are not necessarily chosen because they have a detailed knowledge of museum operations.

The Sjohistoriska Museum follows a rather logical pattern of administrative organization which is centered around the two major areas of marine activity it seeks to portray. Actually, the work of the museum is shared by five departments, but the Royal Navy and Merchant Navy Departments are the two largest and are responsible for preparing and maintaining the exhibits. They also are the only ones headed by curators. Thus the Sjohistoriska Museum has chosen to establish a division of labor mainly along the lines of subject matter rather than in accordance with some general classification of objects. The library and the archives are administered as separate units and organizationally may be considered departments. Their principal function is, of course, to provide a repository for the museum's extensive collection of documents and records and to furnish research assistance to the curatorial staff and other specialists who are encouraged to use their services. The fifth principal division is the Conservation Department which includes the workshop and laboratories. Overall supervision of museum activities is provided by its director, a naval historian, who is assisted by a small personal staff.

In 1958 the Sjohistoriska Museum employed 42 people. This figure is a bit surprising if one recalls that the work of the somewhat larger Armémuseum in Stockholm is carried by a staff of only 10 full-time employees. Actually, the Sjohistoriska Museum has a permanent staff of 23 persons, but this group is normally augmented by an additional 20 to 25 who work on a temporary basis. The Swedish Government provides the services of these people at no cost to the museum. The arrangement is unique, for the group is usually composed of regular government employees who have lost their jobs in other agencies owing to circumstances beyond their control. To prevent them from losing their status as public employees, the government sends them to the museum to work pending relocation in another permanent job or to await retirement in the case of older civil servants. They work wherever the museum can profitably use their skills.

The professional staff of the museum is limited to three persons who carry the rank of curator. These are the director and the chiefs of the two principal departments. Both the library and archives have perhaps the largest staff of any European military museum. The librarian and the archivist both have two full-time assistants. A number of the temporary employees also work in these divisions. The workshop has five permanently employed technicians; four of these are modelmakers. The four guards at the museum also work part time in the shop as painters and carpenters. Two employees serve in the office of the director. One is his secretary who keeps the museum's administrative records, and the other handles most fiscal transactions. Completing the permanent staff are a maintenance man and two janitors.

As a public institution the Sjohistoriska Museum obtains the bulk of its operating funds by governmental appropriation. Of the museum expenditures provided for in the operating budget, salaries account for almost 90 percent of the money. The remaining funds are spent for supplies and furnishings and for conservation of specimens. The government owns the property and assumes all maintenance costs and utility bills. This obligation is rated at about one-quarter of the museum's regular budget. The museum also depends to a considerable extent upon financial support from private citizens. Much of the giving comes from the members of the Association of Friends or is stimulated by their program. However, assistance from other private donors is likewise enlisted, and the director of the Sjohistoriska Museum spends a portion of his time in public-relations activities and fund raising. Whenever a new exhibition is prepared or objects are added to the collections, the museum has to summon the aid of the association and other donors. The government also makes a small supplemental appropriation for these purposes, but the major share of the needed money comes from private sources. Part of the time the museum charges an admission fee. The money received is used to pay salaries for extra guards or to pay for the museum's advertising program. The money needed to finance museum publications—the yearbook, catalogue, and pictures—is obtained from a private donor, but a revolving fund plan prevents its complete depletion. All money obtained from the sale of publications is returned to the fund to apply against further printing costs. A combination of public and private financing for the museum has worked out most satisfactorily, but this has been possible because public interest in the institution has always been high and has been buttressed by a willingness to support the institution with gifts and bequests.

The relationship the Sjohistoriska Museum maintains with the Swedish Royal Navy is similar to that existing between the Armémuseum and its patron service. The navy's official interest in museum activities is monitored by its representative on the board of directors, but there is little indication that the Royal Navy or its companion service, the Merchant Navy, places any formal demands upon the museum. The Royal Navy contributes objects upon request and notifies the museum if it plans to dispose of any surplus equipment. It has its own small collection of antiquities which it prefers to keep intact, but it shows no great reluctance to part with other items. The navy has informed the Sjohistoriska Museum that if asked it is ready to help in any way, except financial, but the museum normally limits its request to those matters which are beyond its own competence.

The beautiful setting which the Sjohistoriska Museum enjoys has been obtained at the cost of a central location. It is situated at the eastern edge of Stockholm, in a general section known as Ladugards Garden, on the northern bank of a body of water called the Djurgardsbrunnsviken. A site near the water while adding to the attractiveness of the surroundings is presumably psychologically advantageous for a naval museum. Immediately behind the building are two other museums. The Sjohistoriska Museum probably feels the competition for visitors offered by these as well as the other fine museums in the city. Nevertheless, the number who come to view its collections averages 40,000 a year, considerably more than the attendance at the Armémuseum, which does enjoy the advantage of a central location. The difference in number of visitors may be due in part to the broader scope of the Sjohistoriska Museum's collections.

MUSEUM FÜR DEUTSCHE GESCHICHTE
Berlin, Germany

Prior to World War II one of the better-known European military museums stood in Berlin on Unter den Linden at the place where this famous street bridges the River Spree, a point now in the East Sector of the city. In those days the museum was called the Zeughaus or Armory. During the closing weeks of the war, the area around the Zeughaus was the scene of extremely bitter fighting. The building was severely damaged, and it ceased to exist as a museum for a number of years. What became of all its collections is not known, but it is presumed they were saved and that some are now on display in the Zeughaus's successor.

The Russians and the Soviet-dominated East Germans have made relatively little progress in rebuilding their sector of the city. With some exceptions, East Berlin is still a shambles, and the awesome ruins of war's devastation are quite often the rule rather than the exception. Only a few buildings have been rebuilt on Unter den Linden, which is somewhat surprising in view of the prominence attached to this street as the heart of old Berlin. A visitor to the East Sector may well be amazed to learn that of the few buildings restored in this area one is the Zeughaus. Further, it has resumed its function as an arms museum under Communist auspices and has been renamed the Museum für Deutsche Geschichte, or Museum for German History.

The building is not completely restored, but sufficient space had been opened by 1958 to permit a sizable display of artifacts. The public could visit one wing on the ground floor and approximately two and one-half wings of exhibits on the second floor. There also appeared to be others in preparation on the second floor. The remaining portions of the rectangular building were in various stages of renovation and repair.

The East German authorities must apparently attach much significance to the museum since they have devoted sufficient resources from seemingly more pressing needs to permit its reconstruction and operation as a public service. It should be recalled that the Soviets place considerable stress on cultural and educational activities and for this reason encourage the satellite nations to reopen their museums, theaters, opera houses, and universities. Probably a more compelling reason for reopening the Zeughaus is that Communist authorities believe this museum has an important educative function to perform in an area that is continually the scene of intense Marxist tutelage. East Germany has given evidence of political instability in the past, and it is reasonable to presume that Communist leaders seek to use every available resource to drive their message across and to persuade the citizens of East Germany that the new Communist way of life is the correct one. Because a military museum traditionally treats the subject of war, it forms a natural vehicle for giving visual expression to the current Communist themes on militarism and peace. Hence, the creation of the new Museum für Deutsche Geschichte has a certain logic in the present pattern of Communist activity.

The fundamental ideological conflict between Communist society and that of the democratic free world is fully reflected in their respective cultural and educational institutions. While the role of the military museum may be essentially the same for each, in that both interpret and portray military history, the underlying purpose and the basic approach taken are considerably different. For this reason, a Communist-oriented military museum could have virtually no validity as a prototype for an American armed forces museum. This is particularly true in the specific areas of fundamental philosophy, administration, finance, and to some extent museum programs and public services.

There are, however, two reasons which make the study of such a museum a worthwhile project. First, there is value in learning something about Soviet military museology and the contribution it is expected to make toward the achievement of Communist objectives. It is an additional facet of Soviet activity which should not be overlooked in our quest for comprehensive knowledge about the Communist system. Second, the techniques of display employed and the quality of specimens, together with their state of preservation, are always of interest to the student of museums, regardless of location. In the course of this study it was impossible to visit a military museum in the Soviet Union. However, the Museum für Deutsche Geschichte in East Berlin fully demonstrates the impact of Soviet influence and can be used to illustrate a Soviet prototype.

Some Soviet attitudes regarding the legitimate function of a military museum have been made a matter of record and underscore the propagandistic nature of the venture. These same attitudes are expressed in the published catalogue of the Museum für Deutsche Geschichte under the title "Waffen und Uniformen in der Geschichte," or "Weapons and Uniforms in History," and are also obvious even to the casual viewer of the exhibition. Only a few Soviet views are discussed here, but they clearly demonstrate what is expected of a Communist military museum, such as the new display in the reconstructed Berlin Zeughaus.

Soviet views on all subjects are presented in the prevailing acceptable terms of Marxist-Leninist doctrine. Hence, it is not surprising to discover that a discourse on Soviet military museums is fully interspersed with such semantics. For example, the Marxist slavish devotion to science is seen in Soviet insistence that their military museums are scientific research and informative institutions. The description of museum work as "scientific" occurs with repetitive monotony and gives the impression that this should be accepted as irrefutable proof of the institution's legitimacy. In an address on May 24, 1957, before the first Congress of Museums of Arms and Military Equipment held in Copenhagen, Denmark, Col. J. I. Vostokov of the Scientific-Methodical Council for the Work at the Arsenal Museum, Moscow, stressed that the real impetus for scientific effort in Soviet museums came after the Bolsheviks attained power in 1917, thus permitting them to become instruments of public education, Soviet style. In the best of Marxist phraseology, Vostokov described this historical transition in these words:

> ... the organization for the museums on scientific lines became especially active after the great Socialist October Revolution, a fact that was bound up with the measures of V. I. Lenin and the Soviet Government radically to improve the museums and bring them into closer contact with the broad masses of the people.

Overtones of the class struggle, Russian nationalism, and the peaceful intentions claimed for Soviet foreign policy all show up in the description of military museum activity. In one instance such museums are depicted as "inseparably bound up with the heroic history throughout the centuries of the fights of the Soviet peoples for their freedom and independence." Such an attitude permits the selective display with appropriate Marxist interpretations of any incident in Russian military history which portrays the masses struggling under Czarist or other leadership for national unity or "liberation from their oppressors." One may question the degree of objectivity attained by military historians in a Marxist environment, but we should be reminded that to the dedicated the application of the science of Marxism to any problem is to attain the ultimate in objectivity.

No one should doubt the truth contained in the Soviet claim that their military museums are educational institutions. Not only are they instruments for informing their viewers how they should interpret the past as good disciples of Marxian Socialism, but also they present graphically the objectives of current Soviet foreign policy. A strong national defense coupled with constant protestations of peaceful intentions throughout the world is the recurring theme of Soviet propaganda. Every agency of public information at the Soviet's disposal at home or in the satellite is marshaled in behalf of this line. The part Soviet military museums are expected to play in this effort is evident in two statements from Colonel Vostokov's 1957 address in Copenhagen:

> It must be emphasized that the displays of the Soviet military museums have solely educational and informatory purposes, and that they reflect the just goals to which the Soviet armed forces are devoting themselves—to defend the Socialist fatherland. In full accordance with the peaceful foreign policy of the Soviet Union, our museums have not and cannot have any militaristic purposes at all. ...

To this he added:

> In their activities the military museums are governed by the principles of Soviet museological knowledge, the most important part of which is an intimate connection between the whole system of collecting and displaying, and the tasks that aim at educating the people in a spirit of love and esteem of the best traditions of their country—tasks which are in conformity with the peaceful communistic construction.

Inferentially, and perhaps not without some justification, Colonel Vostokov attacked the work of military museums in non-Communist countries on two points, and implied thereby that these criticisms could not be leveled at their Soviet counterparts. In one instance he charged that there is an evident indifference to the educational significance of the museum which, in practice, means that displays are not made available to the "broad masses." The result of not having the museums serve the general public instead of the specialist or highly interested viewer is to create a "rarities museum" where the specimens are displayed in some formal arrangement, presumably outside the context of military history.

The other criticism Vostokov chose to make was directed at some techniques employed in display. In this case, he claimed that an eagerness to stress some particular phenomenon or display results "in the crowding out of actual museum specimens in favor of all kinds of auxiliary material without actual museological value." Presumably he was attacking the use of certain types of staging props which detract from the weapons being displayed. He pointed out that Soviet museums build displays around the weapons themselves and that all ancillary materials are used to demonstrate how the weapons were employed and the consequences of their use. He described Soviet displays with these words:

> ... the museum officials endeavor to show the development of the weapons as the result of human activity in the concrete historic situation, in connection with the productive forces, the technical development, and it is at the same time endeavored to make clear how the weapons were used in battle, and the consequences the uses of the weapons concerned had, their influence on the composition, organization, and practice of the armies.

If Soviet domination of its satellites is as complete as is reputed, it may be anticipated that the Museum für Deutsche Geschichte, as a case in point, would reflect the full impact of

Soviet museology. This quickly becomes apparent after examining the subject matter of the museum exhibits and reading the labels and other instructional materials.

In the opening paragraph of its catalogue, the museum announces that its exhibits are helping to formulate a new picture of German history. The old Zeughaus museum is attacked as having been a showcase for Prussian-German militarism and imperialism and one which reflected the bankruptcy of the old system of Hitler's Germany. The catalogue then announces to its reader that imperialism and militarism have been cast aside in the German People's Republic and the Museum für Deutsche Geschichte has risen on the site of the old Zeughaus as a symbol of the recent social revolution in Germany. Finally, in the finest of Kremlin-inspired phrasing, the writer of the catalogue proclaims, "We regard it our duty with our museum resources to take a stand against war and for permanent peace. Our exhibition serves to this end." Thus, this museum takes its place as a propaganda agency in behalf of the Soviet peace offensive.

The Museum für Deutsche Geschichte achieves an identity of interest with its Soviet counterparts in virtually every field. The historical subject matter in most instances is, of course, different. The interpretations are not, for they are drawn fully within the Marxist frame of reference. Both seek to inspire two kinds of action—defense of the fatherland and a readiness to do battle against the forces arrayed against socialism. The objective of the new museum in East Berlin in stimulating such action is stated most succinctly: "May the exhibition provide the knowledge of the need to battle against imperialist aggression and to contribute to the defense of the German People's Republic."

The Marxist explanation of war is intricately woven into the exhibits for all historical periods covered in the museum displays. This means that the weapons and panoply of warfare are carefully related to some period of social evolution that is regarded as significant in terms of the class struggle or that appears to lend historical evidence to the main props of the Marxist thesis. It is perhaps quite proper to suggest here that these weapons merely offer the vehicle through which Communist dogma can be presented in interesting and palatable form. At any rate, they are exhibited with a running commentary describing the economic, political, and social conditions for the age in which they were used. The museum catalogue explains the contextual approach to its exhibits and the services provided thereby for the visitor as follows:

> ... The pervading theme points to the origin of wars within the social order of the time, their character and the motives on which they were based. For example, the relation between productive capacity, social organization and the organization of the army as seen in its development, tactics, ideology and combat strength are shown. Thus, we instruct the visitor as to how he can assess war and the problems of war in accordance with the class struggle, differentiate between just and unjust wars, and from these draw his lessons for the present time.

This statement from the museum catalogue is almost identical with Vostokov's description of Soviet military museum displays as given in his speech at Copenhagen.

The type message the Museum für Deutsche Geschichte conveys to its viewers is transparently obvious from the selection of the subject matter of its displays and the abundance of Marxist phraseology employed in its explanatory material. In addition to one display of ancient weapons, the museum gives a general coverage of military history from the 5th century through World War I. However, certain events are emphasized which have the maximum propaganda value. The survey of history is also divided in accordance with Marxist subjective categories. By selecting its subject matter with great care, the museum has been able to achieve a philosophic unity in which no contradictions are permitted to arise to challenge fundamental Marxist premises. The 5th to the 15th centuries are sparsely covered by the Museum, but this is not surprising in view of the very few military specimens which have been preserved from that age and because Marxist interpretations of history first achieved the level of a comprehensive system with the explanation it offered for the social institutions of the latter days of feudalism.

The museum divides its exhibits into five major chronological groupings. The first is entitled "Weapons in Ancient Society" and presents specimens from the Neolithic, Bronze, and Iron Ages. Marxian interpretations are held to a minimum as they would hardly be pertinent to those periods. The second major division has the name "Warfare in Feudal Society" and embraces historical references from the 5th to the 18th century. The range of subject matter is broad and includes displays of feudal knives and armor, one on the German Peasant War of 1525 which is explained strictly in terms of the class struggle, an exhibit of the Thirty Years War, and a description of the Prussian Army of the 17th and 18th centuries. Obviously, the museum treats some events under the heading of "Warfare in Feudal Society" which most Western historians would not include in such a category. The reason for doing this may be deduced from the manner in which the museum has divided the remainder of its exhibits, for the other three divisions are designedly selected to realize the most forceful presentation of Marxist views.

The French Revolution and most of the 19th-century military history pertinent to Germany are presented under the general heading "Army Matters in Capitalism." The impact of the French Revolution on the development of military forces is seen in the beginning of national citizen armies. The Napoleonic Wars receive scant treatment with the exception of the period 1813-1815, which is designated the "War of Liberation." According to the approved explanation of this event offered by the museum, the German people joined in the struggle against Napoleon in the hope of attaining national unity, personal freedom, and social progress only to have their hope dashed by the sovereign authorities who used the patriotic zeal of the masses to re-establish their own power. With such an explanation, it is hardly surprising that the period after 1815 is labeled one of reaction and the next significant event displayed is the "Citizen-Democratic Revolution, 1848-1849"—a movement in which Karl Marx played a rather important role as a political pamphleteer.

The next division is entitled "Imperialism and War." Two principal exhibits are offered. One deals with "Imperialist War Preparation 1890-1914" and the other covers the first World War. The interpretation of this material is strictly Leninist and patently propagandistic. A considerable number of weapons used during the period are on exhibit, but alongside them are pictures, writings, and other documents of the principal German Marxist leaders of the day. These men are depicted as trying to salvage the nation from the holocaust of a war that was completely detrimental to the German working class. The presentation of this episode of German military history is a badly distorted one to any but a Marxist "scholar."

The final presentation offered to the visitor of the Museum für Deutsche Geschichte is labeled "The Great Socialist October Revolution 1917, the Birth Hour of the Armed Fighting Force of Workers and Peasants." It is frankly commemorative and designed to inspire the viewer with the great significance he should attach to this event. At this point the museum achieves its ultimate as an agency for transmitting Communist dogma. As yet there are no exhibits open which deal with the rise of Hitler and Germany's participation in World War II. Hence, the visitor can depart fully convinced, if he chooses to be, that he has witnessed a graphic presentation of evidence which points to the inevitable triumph of Marxian socialism, the most significant event of which was the victory of the Bolsheviks in the Revolution of October 1917.

Because the Museum für Deutsche Geschichte is housed in the old Zeughaus, it has a very adequate building for its exhibits. Although the wartime damage was somewhat severe, the general lines of this impressive structure remained and restoration could be achieved without significant alteration of the building's original design. The furnishings are also suitable for the museum's needs but are not particularly elaborate, nor do they indicate that any great abundance of resources is available for such purposes.

The museum halls will probably run the full length of each side of the building when reconstruction has been completed. At present two such parallel halls exist on the second floor. They hold the bulk of the collections on exhibit. The first floor hall has several displays of ancient weapons and models of castles and churches prevalent in the early feudal period. It likewise contains the street entrance and sales desk, and at one end is the staircase to the second floor.

The exhibits, particularly those on the second floor, are planned chronologically, so that the visitor enters at the point where he sees military specimens of the 5th through the 12th century and leaves after having viewed the 1917 Bolshevik Revolution display. The desired traffic flow is assured by the sequence of displays and by a number of posted instructions. Because the building restoration is as yet incomplete, there are some apparent exhibit space limitations. This does not appear to have impaired the quality of existing displays. What portions of the collection are not exhibited or what plans exist for filling in the obvious gaps in subject matter are not known. What is shown now is not particularly crowded and has been assembled quite attractively. The two lengthy halls are not broken up by wall partitions, but the exhibits are placed in well-defined units. With certain exceptions, the quality of the exhibits in the Museum für Deutsche Geschichte as seen in design and execution reveals a technical excellence comparable to that which exists in the best military museums in western Europe. This is particularly true in the use of effective staging techniques to focus the viewer's attention on what is deemed significant.

The preservation of the armor, firearms, edge weapons, and uniforms is of good quality and at the level attained in other museums. The museum also has a number of superb dioramas which illustrate particular troop formations and famous battles such as that of Jena in 1806. The miniatures in the dioramas are the two-dimensional type now prevalent in many European museums. The models and dioramas are perhaps the most outstanding items in the museum collections.

Modern techniques have not been utilized in preserving the museum's flags and colors. With few exceptions, the now-abandoned method of using coarse mesh has been employed. They likewise are all displayed in the open air, and for the most part they show an advanced state of deterioration. The uniforms appear to be in a good state of preservation and are now exhibited on hangers suspended on small ropes or heavy cords fastened to the tops of the showcases. This is probably a makeshift solution to the problem of displaying a sizable number in a limited space. An alternative solution would be to employ manikins, but very few are found in the museum. Perhaps sufficient funds are not available to purchase the goodly number needed, or else this is an item of museum equipment which cannot be found in quantity within the Soviet bloc.

A critical examination of the labels used in the Museum für Deutsche Geschichte requires that separate consideration be given to their format and content. The format is generally well executed. The inscriptions are neat and, of course, comparatively new. As in most museums, two types of labels are used. One bears the simplest of identifying data and is employed mostly in relation to particular specimens. The other type is explanatory in nature.

This is the vehicle used so ably to interpret the Marxist message and these labels contain the full range of Communist semantics, slogans, and catch phrases.

Information about the financial base and administration of the museum is considerably limited; however, certain general observations can be made. In view of the function this museum performs in a Communist environment, it is obviously an institution of the state and is publicly financed, although the quantity and allocation of its funds would be difficult to discover. What the museum's precise position in the governmental structure is is not known, but presumably it is within the jurisdiction of a ministry which has general supervision over cultural affairs. It is somewhat doubtful that it has been incorporated in the defense structure.

Information on the number of employees is lacking. The methods employed within the museum and the kind of work performed are little different from those existing elsewhere in Europe. Hence, a link with the German and European museological tradition may easily be presumed. The technical requirements call for the museum staff to include the usual complement of curators and technicians. The political orientation of the professional staff is certainly not in question, for the result of their efforts is transparently visible. The quality of specimen preservation indicates that there are qualified technicians working with the armor, firearms, edge weapons, and uniforms. On the other hand, the state of the flags and colors indicates that the museum does not have a highly qualified technician in this field or else materials for adequate preservation are not available.

The guards employed within the museum during the hours the exhibit is open to the public are all middle-aged women. They are uniformly dressed in black shoes and stockings, black skirts, and dark-green waistcoat-type blouses. Their somber appearance underscores the seriousness of purpose apparently attached to the museum. Most assuredly, they detract from the otherwise favorable physical appearance of the museum.

The Soviet-influenced Museum für Deutsche Geschichte leaves at least one lasting impression upon a visitor from the non-Communist world, for it conclusively demonstrates that the prevailing national political ideology completely guides all museum activity and finds in the museum a most useful form for its expression. The museum clearly demonstrates that within a totalitarian framework it can be molded into a dynamic institution to justify present social, economic, and political developments in terms of a carefully formulated Marxist interpretation of past military history. Regardless of any technical proficiencies which may exist, it is the very lack of free inquiry existing in this museum that condemns it as a useful model for an American armed forces museum.

HEERESGESCHICHTLICHES MUSEUM
Vienna, Austria

During World War II the famed Austrian Army Museum in Vienna sustained such severe damage that almost complete reconstruction was necessary. The two great wings on either side of the domed center section of the exhibition building were badly bombed and damaged beyond repair. However, the central hall remained standing and in good enough condition to be restored. Because this most striking part of the old building could be reclaimed, Austrian officials decided to reconstruct the Army Museum in much of its original architectural style. However, in so doing they determined to correct many defects which had existed previously and to provide new facilities that would meet the needs of a modern museum. Through this decision there has emerged from the rubble of war's destruction perhaps the finest military museum structure in Europe. Certainly it is the newest, and thus it is in marked contrast to most European military museums housed in buildings of considerably earlier vintage.

The Heeresgeschichtliches Museum, its present title, is located among the old army arsenal buildings. The original museum building was constructed between 1850 and 1856, but the Austrian Army Museum first opened its exhibition to the public in 1891. With the reconstruction of the present building, which received its first visitors in June 1955, the Army Museum entered upon an expanded phase. First of all, this new structure contains about three times the space of its predecessor. Greater emphasis is now given to recording important events in Austria's military history than in former days, although a contextual description of these events is held to a minimum. Perhaps the greatest change has occurred in the setting of the museum displays, for the Heeresgeschichtliches Museum must be rated as one of the most impressive and esthetically attractive military museums in Europe.

The title of the museum would indicate that its exhibitions are limited to army objects. For the most part this is true, but in reality the Heeresgeschichtliches Museum assumes the character and functions of a national armed forces museum. Most Austrian military history has been written by the nation's ground forces, and so the great emphasis given to the army by the museum is a natural consequence. The geographic location of the old Austro-Hungarian Empire precluded it from becoming a great naval power, but the Empire's naval history is given appropriate graphical presentation in the museum's exhibits. The air service, however, receives virtually no representation. This is probably the case because the museum's displays do not presently go past World War I, and, indeed, the treatment of this epic is quite limited. Neither does the museum attempt to cover the full sweep of Austria's military history. Rather, its exhibits are restricted to what is called the "later history of Austria," and its chief displays are concentrated around the significant events and personages of the 17th, 18th, and 19th centuries. The period treated coincides with that in which the Austro-Hungarian Empire achieved a period of prominence in international politics, and to a great extent it chronicles the deeds of the Hapsburgs.

The newly reconstructed building which houses the displays is only one of three which the museum presently occupies. It is located inside the courtyard of the arsenal near the main gate. The museum offices and library are in the building which forms part of the entrance to the arsenal grounds. To either side of this building are lengthy colonnades which house many of the museum's fine collection of cannon. The workshops and other service facilities are located in a building which forms part of the eastern edge of the arsenal courtyard. The museum also plans to recondition for exhibition use the building on the west side of the arsenal grounds which is the counterpart of the one occupied by the workshops. The offices and service buildings were also damaged somewhat during World War II. Because of their recent renovation, they are quite adequate for their present purposes. The arsenal grounds

Heeresgeschichtliches Museum, Vienna. The Hall of Military Commanders.

Heeresgeschichtliches Museum, Vienna. The Maria Theresa Hall. Note the period design of the furnishings.

Heeresgeschichtliches Museum, Vienna. The Archduke Charles Hall.

Heeresgeschichtliches Museum, Vienna. The Prince Eugen Hall.

are nicely landscaped, and so the setting for the new museum building is a most attractive one.

The public enters the exhibit building through an impressive central room called the hall of military commanders. This has been built in cryptlike shape with a vaulted ceiling and a number of massive pillars. About these pillars are grouped 56 life-size statues in marble which commemorate Austrian sovereigns and their outstanding military commanders. This memorial to Austria's kings and military heroes establishes the museum's patriotic theme, and the visitor tends to linger within an inspirational atmosphere before beginning his tour of the exhibits.

Visitors may remain on the first floor to see the navy hall and other displays dealing with the more modern periods of Austrian military history, but normally they start their tour on the second floor so as to follow the chronological sequence of the exhibits. There is also a natural tendency to go to the second floor because the staircase is immediately across the room from the front entrance, and a corridor effect is created by two central rows of statues. There is also a floor runner from the entrance to the staircase which automatically directs the visitor's attention to the second floor. The staircase itself is a major artistic achievement. It contains a number of life-size statues and busts of Austrian generals and admirals and its vaulted ceiling is decorated with colorful allegorical frescoes by Karl Rahl.

Entrance to the second-floor exhibits is gained through the most impressive room in the entire museum. This is the cupola-crowned hall of fame, which pays tribute to the memory of the Imperial Austrian Army. The hall has an annex at either side which leads into the two lengthy wings containing the displays. The magnificent frescoes by Karl Blaas which adorn the ceiling and upper sidewalls depict the most important events in Austrian history. Above the arches and doorways around the room are hung a number of flags which are trophies of the major wars in which Austria engaged during the 17th, 18th, and 19th centuries. The only object on the floor of this spacious room is an octagonal glass case in the center which provides a legend of the museum's principal exhibits. At this point, through the use of documents, maps, and a few objects, the museum recalls to the mind of the viewer Austria's major military achievements and forecasts what he may expect to see during his tour.

It is possible to see the full length of the museum's two wings from the hall of fame. Nevertheless, each is broken into a number of separate halls by means of archways and curved ceilings. Often the ceilings are frescoed and in some instances contain reproductions of Austrian coats-of-arms and other heraldry. On both floors of the museum, the rooms are normally named after a major figure in later Austrian history. For example, halls are designated for Maria Theresa, Archduke Charles, Prince Eugen, Emperor Franz Joseph, and the great commander Radetzky. Exceptions to this pattern are found in the navy hall and the Sarajevo hall, which commemorate in one instance an entire service and in the other an event which has had a major impact on modern Austrian history.

All exhibits reflect the museum's apparent attachment to the fine arts. Each room has been designed to make it as attractive as possible and to insure that it reflects the appropriate time period. This is achieved through the display of significant and attractive objects; the use of symmetry in the positioning of cases, pictures, and flags; and by creating furnishings which are in the contemporary design of the epic treated. This latter technique greatly helps to produce a unity within each major display. The entire museum is kept immaculate, and the floors and cases give evidence of frequent and recent polish.

Because the museum's exhibition space was increased threefold when the building was restored, there appears to be more than ample room for existing displays. The staff also plans to add to its present exhibits by opening new halls which will bring the graphic presentation of Austrian military history up to date. However, there is no intent to reclaim for this purpose exhibit space now in use. Rather, additional space in the exhibit building will be opened and another building available to the museum will be renovated and displays placed there. The museum is indeed fortunate that it can meet its needs in this manner. This is largely possible because the Austrian Government regards the nation's museums, particularly those in Vienna, as a major national resource. It thus lends its support to the Heeresgeschichtliches Museum staff's objective of achieving a high standard of excellence.

Throughout the museum the viewer gains an impression of spaciousness. There are no cluttered aisles flanked by a multitude of cases, each packed with a great quantity of objects. Instead, each room contains very few cases, and these are widely separated. Only a few items are displayed in each, and they are placed so that every object may arrest the viewer's attention. Often a case will contain a single object—a uniformed manikin, a ship model, a weapon, or a famous trophy. Only the striking and significant items in the collection are displayed. The remainder are retained in the study collections for use by the staff and interested students.

Each exhibit normally consists of a few weapons, uniforms, medals, busts, flags, maps, and pictures and some other significant and appropriate objects. For the most part, the displays are permanent, but they are supplemented from time to time with new acquisitions. A hall has been set aside for special exhibitions which are changed two or three times a year. With regard to the permanent character of its displays and its program of special presentations, the Heeresgeschichtliches Museum's policy is similar to that of most museums.

Because the exhibits are organized around significant figures in Austrian history, there is seemingly little attention given to the anonymous soldier in the ranks. The personal objects of sovereigns and commanders are on display together with some representative arms and uniforms of each period. The viewer is thus attracted to the mementos of the hero and may possibly lose sight of the contributions of the average soldier. This method of exhibition somewhat limits the museum in its capacity to instruct the viewer in the basics of Austrian military history or to catalogue a flow of events which will enable him to view the objects within a rather fully developed historical context. Significant events, of course, are mentioned. However, the visitor is well advised to acquire some fundamental knowledge of Austrian history before arriving at the museum if he is to appreciate fully the significance of what he sees.

Exclusive of the personal mementos of military heroes, the museum's inventory has approximately 25,000 individual items. The weapons collection includes 2,200 small arms; 2,600 sabers, swords, and daggers; 380 pieces of heavy ordnance; and a few items of air equipment; but no motorized artillery or tanks. It possesses 1,700 flags, standards, and pennants; 2,000 medals;

2,800 uniforms; and 2,700 head pieces which are counted separately. There are also a few ship and cannon models. Completing the inventory is a rather extensive art collection consisting of about 4,000 oil paintings and 6,700 prints and drawings. The finest canvases are featured to good advantage in every hall, and many of the prints, flags, and photographs have been mounted in revolving stands and placed in the various exhibits to which they relate.

The Heeresgeschichtliches Museum currently has about 59,000 square feet allocated for interior exhibit. This space will increase by several thousand square feet with the enlargement of the navy hall and the completion of a new sizable artillery hall, both of which are in process. The large colonnades which now house much of the cannon exhibit provide a protected exterior exhibit space of 48,000 square feet. In classifying the museum's space utilization, the staff makes no differentiation between the area assigned to ready reference materials and that given over to other forms of storage, but for all categories of storage there is an area of 34,000 square feet. The large and well-equipped workshops occupy almost 14,500 square feet, the offices 4,500, the library and study rooms 2,400, and a refreshment room 1,200, and there is an auditorium occupying about 2,600 square feet.

It is quite evident that all employees of the museum take much pride in the present attainment of their institution, for most have participated in the difficult work of restoration. They had to repair and preserve many objects in the collection before these could be placed in the exhibits. Almost the entire inventory of museum furnishings was built in the workshops. The professional staff does not view the museum merely as a repository of significant historical objects, but rather they think of it in terms of a major center for the study of military history, one which is closely allied with other higher institutions of learning and increasingly contributing the fruits of its own creative endeavor. The museum thus provides an academic atmosphere through the activities of its curators who are primarily military historians with particular fields of specialization.

Because the museum's educational functions receive such strong emphasis, the curatorial staff has devoted much energy to research. This research endeavor is manifest in several programs. By far the greatest effort has been concentrated on accumulating data about objects in the museum collections and the circumstances of their use. Much of the information gathered is now incorporated in the explanatory material provided in the exhibits, but these data also provide an excellent source of collateral material for use in future special research projects. The staff also spends some time in answering questions addressed to the museum by the public. A great number of inquiries are received which can be dealt with by a simple answer, but these are not recorded for statistical purposes. The ones requiring considerable study are listed under the title "Scientific Questions" and now average about 250 a year.

The third type of research performed by the staff is in response to assistance requested by the University of Vienna. Normally the university asks for data related to its own specialized projects rather than for finished studies. In essence the university is doing little more than drawing on the resources of its own staff since several of the curators serve on its faculty. This type of working relationship between the museum and the university is a long-standing Austrian tradition wherein the nation's universities draw on museums, art galleries, libraries, archives, and other learned or cultural institutions for a considerable number of their faculty members.

In addition to this extensive program of service to others, the staff has been able to carry on additional independent research. The members have collaborated on preparation of a new catalogue, a history of the museum, and a monograph discussing the cultural role of the Austro-Hungarian Army during the 19th century. There have also been other publications and more are contemplated. Although the museum budget does not specifically provide for separate funds for research, some money is available for travel in connection with staff research activities and for the publication of studies, monographs, and other papers. The museum also encourages visiting scholars to use its facilities and has provided living quarters for a visiting researcher while he uses the museum's archives and reference collections.

The officials of the Heeresgeschichtliches Museum have attempted to obtain strong public support for their institution.

They have pursued this objective in several ways. First, the museum has been made as impressive and as attractive as possible. This is most desirable in view of the competition offered by many other excellent museums in Vienna. Its patriotic theme, the presence of the relics of Austrian national and military heroes, and the beauty of the exhibit building combine to make the museum a prime public attraction. Special lectures are offered every Sunday in one of the museum halls and public attendance is cordially invited. An expanded lecture program is now contemplated with the completion of the auditorium. Motion pictures will also be shown. Members of the curatorial staff give lectures to special groups upon invitation. When they can be scheduled, guided tours are provided for which there is usually a small charge. A number of these tours are conducted for school groups and units of the Austrian Army. Finally, provision is also made for the visitor's comfort. A number of the halls have seats where the viewer may rest, and there is a sizable refreshment room on the ground floor.

Members of the staff are willing to offer their services as consultants to amateur military collectors. Some of the technicians will restore objects upon request, but a fee is charged for this type of assistance. Identification of specimens and providing pertinent information are the routine services which the Heeresgeschichtliches Museum furnishes in common with most military museums. It also adheres to the common practice of not lending any portion of its collections to the general public. However, this courtesy is extended to other museums in Vienna and elsewhere throughout the country, as well as to other institutions which require military objects for display purposes during special events. There are several other military museums in Austria, but most are private institutions and the Heeresgeschichtliches Museum normally does not work closely with them, nor assuredly does it exercise any control over their activities.

The excellent reputation of the Viennese Army History Museum is now firmly established throughout Europe. In part that reputation rests upon the architectural splendor of a new building. Because it is ornate, the museum appears somewhat opulent, a characteristic denied to many other continental museums which have not been accorded such a degree of governmental beneficence. Although it must share many of the Hapsburg mementos with the Kunsthistorisches Museum, the quality of its artifacts and the fine appointment of its exhibits contribute further to the museum's position of excellence. But to a great extent the Heeresgeschichtliches Museum bases its claim to high regard on the quality of its staff and the contributions which its members make to the field of military museology. In addition to the scholarly endeavor of its curators, the museum is widely noted for its restoration of flags and paintings. It is doubtful that any other military museum in Europe surpasses it in either category, and very few are able to offer it serious competition.

Like many of the great museums in Austria, the Heeresgeschichtliches Museum is a public enterprise. It is attached to the Ministry of Defense for administrative purposes. The pattern of control exercised by the ministry is similar to that found in many other European countries, for it uses its authority very sparingly and grants the museum virtual autonomy in the conduct of its routine operations. The museum authorities report directly to one of the highest officials in the ministry who is equivalent to a permanent undersecretary and regarded chiefly as a point of contact on matters which require official sanction. The chief method of general supervision which the government employs is, of course, fiscal. The museum budget must be approved and incorporated into that for the entire defense establishment, and operating funds are acquired through legislative appropriation. The government must also give approval to staff appointments, and it completes the discharge of its supervisory responsibilities by requiring two systems of reporting. One is performed by the museum director through a monthly and annual report of the institution's activities. The other is an audit of the museum accounts conducted by the government. Through these means, the government is able to perform something of a permanent inspection, but most of this is accomplished by paper work.

The Heeresgeschichtliches Museum has no board of directors, advisory committee, or other policy-making body apart from its own officials. The responsibility for all operational policy determinations is lodged with the director. The advice he receives on such matters emanates from within the professional staff when he requests it.

The museum's organization is a model of administrative simplicity. There are no formal departments, but each member of the professional staff is a specialist in some museum field of interest, i.e., swords, guns, uniforms, and heraldry. Thus, a natural division of labor results without the requirement for administrative formalization. The director gives instructions to each curator and the chief technicians and provides immediate supervision to all those in a clerical capacity, or those who perform administrative services for the museum. His span of control if fairly broad, but organizational unity and personalization of direction also result.

The personnel policies in the museum also re-enforce the effort to attain a sense of unity. The need for professionalization is stressed at all levels, and each individual is urged to regard himself an important contributor to the museum program. He is encouraged to believe that any diminution of his efforts will lessen the museum's capacity for high achievement. The development of a team concept and a policy of recognition given for technical competence have enabled the museum to achieve much in a relatively short period and to attain an established reputation for high quality.

There are normally 67 persons employed at the Heeresgeschichtliches Museum. Five of these, including the director, are designated as academic officials with the rank of curator. The chief restorer of pictures is also a highly trained academician and is accorded a place on the professional staff. The curators and director are all historians and have a doctor of philosophy degree or its equivalent. Three of them are professors at the University of Vienna, and one serves on the faculty of the Technical High School. The teaching requirement for each is four to five hours a week. None have had a professional career in the armed forces, although some may have seen service during the war.

In addition to the director and members of the professional group, the museum has a large number of technicians with widely diversified specialities. There are three working on the restoration of pictures, six embroiderers who restore battle flags and standards, seven carpenters, two tailors, two locksmiths, two master weapons conservators, two photographers, two librarians, and two book binders. Completing the ranks are seven who perform clerical and other administrative tasks, 19 warders, four night watchmen, three servants, and a waitress.

The Heeresgeschichtliches Museum budget is quite similar to those of other European military museums. The specific allocation of funds within the annual outlay is not known, but the budget covers salaries, purchases of supplies, conservation of materials, and acquisition of new objects. The Austrian Government normally absorbs all maintenance and upkeep costs, as well as every capital improvement to the property. The entire cost of rebuilding the museum after World War II was also met by public finance. Officials of the Heeresgeschichtliches Museum claim the funds they receive are quite adequate, and hence they have no pressing financial problems. The splendor of their facilities offers demonstrable proof of this. This fortunate position stems from the attitude the government takes regarding the country's museums. Because these institutions are officially viewed as national resources and a prime tourist attraction, they possess a long-range educational and economic value which the Austrian Government deems worthy of investment. It is this strong governmental support which has brought the Heeresgeschichtliches Museum to its present level of national and international prominence. The museum charges a small fee for admission, but the funds collected revert to the national treasury and, to the knowledge of the staff, are not applied specifically to museum expenses. The museum has no private income of any kind, although occasionally interested citizens donate worthy objects to its collections.

Although the Heeresgeschichtliches Museum is a prime center for the study and commemoration of Austrian military history, it does not maintain a very close official relationship to the nation's armed forces. The museum welcomes the support of the army and offers some instruction to its units, but the army apparently exercises no influence over museum policies. It will perform some services for the museum if requested, but this is done very infrequently. The army donates uniforms but not weapons. There is no indication that this lack of close collaboration with the army hampers museum operations in any respect.

At present the Heeresgeschichtliches Museum is not so fortunate in its location as is the Kunsthistorisches Museum and other celebrated Viennese museums situated along the city's famous ring or within the inner city and thus in the central tourist area. However, it is readily accessible to anyone who desires to visit it and can be reached rather quickly from the heart of the city, provided public transportation is not required. The museum is only a few blocks from Belvedere Palace, which was once the residence of Prince Eugen, and is today one of Vienna's major tourist attractions. The approach to the museum is through a small park known as the "Schweizer Garten." Future plans call for additional museum facilities in this park, and the potential of the area as an attraction for visitors will be increased. When this occurs perhaps the number of visitors to the Heeresgeschichtliches Museum will considerably increase beyond its present average of 400,000 a year, which is the largest number recorded for a military museum in western Europe.

MUSEO DEL EJERCITO
Madrid, Spain

The rich military history of Spain would require many volumes to recount, for it embraces a period of over 2,000 years. There are many chapters of abiding interest which begin with the age when Spain was included as a major colony within the Roman Empire. Others tell of the centuries when the Moors extended their domination over the country and left a permanent impact upon Spanish culture. For the student of later history, the names of Hernan Cortes, Francisco Pizarro, Cabeza de Vaca, and Vasco de Balboa excite the imagination. The exploits of these conquistadors who were a combination of explorer and soldier of fortune in the service of the king extended the power of Spain to the New World and marked the beginning of an era when Spain challenged the might of England on the seas and rose to a position of major military prominence on the European continent. In more recent days, the military history of Spain has been recorded in the devastation wrought to the land by civil war. All of these epics offer a challenge to the museologist as appropriate and fascinating subjects for graphic portrayal. Unfortunately, the earlier periods must largely be excepted from museum treatment, for the preservation of objects in use prior to the 15th century is almost nonexistent.

Some of the antiquities now comprising part of the Museo del Ejercito's prized possessions probably were first assembled as private collections by officers and soldiers who derived personal pleasure from gathering and displaying military souvenirs. These are the sources to which many military museums trace their origin. The date which the Museo del Ejercito (Army Museum) came into existence in something of its present form is not known, but part of its collections have been public property and on display at one place or another in Madrid since 1803. Although the museum now has a continuity of operation extending for more than a century and a half and records many of Spain's most notable past military achievements, there is still the firm touch of the present. No visitor can tour its halls without being reminded that the Civil War of 20 years ago has made a tremendous impact upon Spain, nor can he fail to discern something of the power the Caudillo, Gen. Francisco Franco, now wields over national affairs.

The grandeur of the past is reflected in the building which houses the Museo del Ejercito. During the years 1631 and 1632, King Philip IV constructed his beautiful Palacio del Buen Retiro, which continued to serve as the center of much governmental activity until the close of the 18th century. Most of the original building has now been removed, but one wing still remains, and this has been appropriated for use by the museum. The interior is quite ornate, and many of the original decorations and appointments have been preserved or restored. The royal staircase has been retained, and the original room arrangement has been little altered. Beyond this, there is little evidence that this building was once a royal palace, although the historical importance of the building is fairly well known, at least in Madrid. Thus, some visitors probably come to see the museum more because of their interest in the building's past history than through a desire to view its military objects.

The edifice is rectangular and is constructed of a combination of stone and brick. It is three stories high, although at either end there is an intermediate floor between the second and third stories. The only external distinctive feature is a short steeple at one corner. From the standpoint of modern architecture and its present surroundings, the structure is not particularly impressive. However, when it was an integral part of the original palace and surrounded by attractively landscaped grounds, it probably rated as one of the most remarkable structures in the Madrid of a bygone era.

The Museo del Ejercito is entered from a narrow esplanade which runs the length of one side of the building. The first object to be seen in the entrance hall is a bronze model of the Alcazar of Toledo. Through this symbol of Franco's rise to power, the visitor's first contact with the museum exhibits is the most recent chapter in Spanish military history. Both the entrance hall and the adjoining royal staircase by which the visitor goes to the second floor are fitted with richly carved ceilings. A more recent addition to the royal staircase which highlights the military character of the building is a balustrade made entirely of small guns. The visitor is encouraged to begin his tour on the second floor. The catalogue which is sold at the main entrance is most helpful in directing the visitor through the rooms with the starting point the Queen's Little Hall at the head of the royal staircase.

The arrangement of rooms on each floor is almost the same. Each has a large hall which runs almost the full length of the building. On the second and third floors this hall has been partitioned to make three equal-sized rooms, but on the ground floor it has not been subdivided. There are additional rooms at either end of the major hall on all floors, and on the ground and first stories there are several rooms adjacent to it along one side. Thus the floor arrangement features a large central hall surrounded on three sides by a series of smaller rooms. These are not interconnected, however, and must usually be entered from the main hall. As described in the catalogue, the visitor's tour is planned so that he will not have to retrace his steps, even though he enters and leaves the central hall on each floor several times.

In general, the exhibits have been prepared as a combination chronological-topical presentation with major emphasis given to the topical elements. It is impossible to trace the time sweep of Spanish military history in any series of halls, although at one or two points in the museum, adjacent halls treat contiguous periods. Rather, the plan appears to be one of grouping the weapons and related materials of various components of the Spanish Army in the principal halls and placing other subjects treated where it seemed most convenient. For example, the Arab hall displaying trophies of early Spanish conquests in Africa is entered from one of the infantry halls and is immediately adjacent to the room commemorating Franco's victory in the Spanish Civil War only two decades ago. Thus, the viewer is constantly shifting from one time reference to another and may have some difficulty in integrating what he sees into some context of events, unless he has a reasonably good acquaintance with Spanish history. The exhibits have been arranged so that the collection of artillery is on the ground floor, the small arms and objects used by the cavalry on the second floor, and on the top floor relics of the Napoleonic period and the contributions of the army engineers and quartermaster corps.

The museum contains many excellent objects in its collections and has a number of features which should be noted briefly to convey a more comprehensive picture of its scope. First, virtually every major element of the Spanish military establishment has an exhibit hall of its own or at least a special display. In addition to those which have been mentioned, there are halls for the Coast Artillery, the Medical Corps, and the Civil Guard. The latter two are comparatively recent additions to the museum exhibits and were both opened by General Franco in formal ceremonies. The Museo del Ejercito contains one feature not found in most of the other military museums in Europe, for it has a special hall commemorating the contributions women have made to the nation's military history. The heroines of Spain are duly memorialized by a series of portraits, miniatures, and other objects. Most of the women included in the group distinguished themselves in behalf of the Franco forces in the Civil War and in all cases were killed in combat or martyred after capture.

Models are used to excellent advantage in a number of the displays. The artillery collection is particularly noteworthy and features various types of recent and earlier weapons. Many of the models have the full carriage mount, and some of the 18th and 19th century cannon model assemblies also include a team of horses and the ammunition carriage. This series has considerable educational merit as well as interest appeal for the large number of visitors who invariably are intrigued by a display of military models. The museum also uses models and full-scale objects to illustrate technical research and the fabrication of weapons. Bench models of fabricating machinery are used in conjunction with actual parts to demonstrate the assembling of various weapon components. Perhaps the best use of models is found in the two engineers' halls. In one room, topographical models are employed to show such engineering achievements as trench construction, mine warfare, and types of fortifications. The other hall contains a set of models illustrating the types of bridge construction the army engineers have employed.

One other hall of models provides a particular delight to children and has been named the children's hall. This room contains almost 20,000 lead soldiers placed upon five tables and several shelves. For the most part they are displayed in troop formations and represent such armies as that of Napoleon's, the Spanish Army at the end of the 19th century, and soldiers who fought in World War I. These models also are used to trace the history of uniforms used by various corps in the Spanish Army. The Museo del Ejercito has one of the largest collections of model soldiers found in any military museum, and this provides it with one of its major entertainment assets.

Because the museum building is now over 300 years old and was adorned in a manner worthy of a 17th-century palace, the Museo del Ejercito sustains throughout an aura of a bygone age of grandeur. To have remodeled it with the objective of modernization would probably have been financially prohibitive and would certainly have deprived the building of much of its historical character. On the other hand, to some persons use of this structure as a military museum might seem a disservice. To those who like to reminisce about past events and reconstruct them, a display of weapons and flags seems out of place in the kingdom hall where the nobles of the Spanish realm comprising the Cortes gathered for their deliberations, and where other ceremonies of state were customarily held. Be that as it may, preservation of the building in much of its original form provides an apt setting for the museum's military antiquities. In such surroundings even references to the Spanish Civil War of the 1930's and control of the state exercised by Franco seem somewhat in the long-established military traditions of the nation.

Most of the rooms are attractive in their ornamentation and appointments. The principal halls excel in these matters, but virtually all rooms contain their share of richly carved tables and cases. Some rooms have been designed with a motif which vividly conveys the general theme of the exhibit. The most striking example of this is the Arab hall, which strongly resembles the interior of the Alhambra and includes various pieces of Arabian style furniture.

Two other halls command the viewer's admiration because of their attractive design. One of these is the hall of the Chapter of the Royal and Military Order of San Fernando. Lining the walls of this solemn hall are large oil portraits of the generals and admirals who are now entitled to wear this, the highest military order in Spain. The other hall commemorates Spanish overseas exploits and features military objects from the many lands touched by Spanish arms and explorations. In introducing the visitor to the overseas hall, the writer of the museum's catalogue modestly proclaims that "the country which discovered a New World, which along with Portugal was the ruler of the old one, and which from the first days of its history went over the earth, cannot present a hall dedicated to its colonial souvenirs, for its museum is all the world, where there is not a piece of it without a Spanish grave." The exhibit has been prepared with obvious pride in Spanish accomplishments of the past.

The exhibits of the Museo del Ejercito generally create the impression that the museum is principally concerned with its custodial function—that essentially it seeks to provide little more than a repository for the souvenirs and relics which are significant to Spanish military history. This does not necessarily depreciate its value as a museum, but it does restrict its role primarily to that of caretaker for the past. It buttresses its custodial function by displaying in great quantities the portraits of military commanders and normally describes each with a single sentence in the catalogue. This notation usually identifies the particular engagement in which the hero met his death but seldom includes the date or any other relevant data. If additional explanatory material were added, an interested visitor who possesses little information might be provided with much useful knowledge about Spanish military history and his visit to the museum made an even more gratifying experience.

Insofar as the Museo del Ejercito regards its primary function as that of preserving and displaying interesting and significant objects used either by the army and key leaders or captured in combat, it is effective. With the exception of a number of its older flags, the artifacts generally are in a good state of preservation. However, many of these older colors may soon be lost to the museum because they are of such old fabric, and thus present a most difficult task to the restorer.

The tendency in the museum is to display a large number of objects. This, of course, means that the visitor must spend a great deal of time in the museum if he wishes to look at the displays in any great detail. Great numbers of swords are displayed together. The detailed designs of the flags and colors are sometimes difficult to see. With few exceptions, they are displayed upright in clusters, and some are high up on the walls.

The remaining objects are well exhibited. This is particularly true for the armor, small arms, artillery, and uniforms, most of which are displayed on manikins. In general, space is efficiently utilized, and in all rooms the visitor can move freely about the showcases and tables. Yet in some of the halls there is considerable clutter, and an impression of crowding results. Such rooms tend to show complete collections or large numbers of objects; additional space is needed if the display is to be more attractive. The museum inventory contains many pictures, most of them portraits of army officers, and these pictures cover the walls of almost every room in a quantity one finds in a portrait gallery. Apart from memorializing the well-known and lesser commanders of the army, these portraits provide an excellent record of Spanish uniforms. Busts and statues of famous personages are also included in several of the halls.

In recent years, the museum has undergone a few changes, resulting in the addition of two or three new halls. No further additions are contemplated, and all the exhibition space in the building is now being used. Hence, the Museo del Ejercito finds it is unable to hold special exhibitions. The basic displays seldom change. New pieces are added on occasion, but the over-all exhibit is permanent.

The inventory of the museum contains a general collection and several smaller ones which have been donated and still retain their identity. In two instances these have their own exhibition rooms. It is difficult, therefore, to estimate the total number of items in the museum, but some indication can be derived from listing the categories of objects and specifying some of the amounts.

The museum notes that its exhibit halls have 98 uniformed manikins on display. These are well distributed among the chief components of the Spanish Army and also include uniforms from Morocco and Japan. The small-arms collection contains almost 900 weapons or components, about 40 machine guns, and 160 pistols and revolvers of various types. The edge weapons include over 300 swords, ceremonial weapons, sabers, and sword blades; 45 knives and daggers; a separate collection of cavalry saber models from almost every country; and a collection of arms from the best Toledo swordmakers. In the category of antiquities, the museum possesses over 100 war axes and clubs; a number of bows, crossbows, pikes, arrows, and slings; and a collection of 15th to 17th century armor displayed on manikins, each equipped with contemporary weapons. The artillery collection contains over 150 cannon, 38 mortars, and 32 howitzers. On display with these are a large number of projectiles ranging from stones to modern artillery shells. In addition, there is the museum's rich collection of models of various types. The Museo del Ejercito has been most fortunate in obtaining the personal relics of famous soldiers and other soldier-collectors and now has about 130 such collections. Added to these are a large quantity of medals and awards, an extensive collection of flags and banners from all the Spanish regiments, and a vast quantity of paintings. A sizable library and a complete set of documents authenticating objects in the collections complete the inventory.

By European standards, the Museo del Ejercito is medium sized and allocates about the same amount of space for interior exhibit as does the Imperial War Museum in London and the Armémuseum in Stockholm. This is approximately 43,000 square feet. The esplanade between the building and the street provides the museum with 8,600 square feet for outside display. Not much of the space is used, however. A statue, two mortars, an antiaircraft gun, and a howitzer stand near the entry steps to the esplanade, and a few guns and mortars have been placed against the walls of the building. An 8-ton Vickers tank is the largest item displayed outside. The museum has some storage space, but the exact area is not known. Judged from the amount of room used for exhibits, offices, and service facilities, space available for storage within the building is rather limited. The library occupies an area of 750 square feet and approximately 1,650 square feet have been allocated for offices. In the way of service facilities, the museum has a workshop, a photographic laboratory, and an exhibit laboratory, all of which are fairly small.

The Museo del Ejercito is perhaps the most closely linked to its patron service of any military museum in Europe. It is a part of the Ministry of the Army and is directly under the control of the Subsecretariat of the Ministry, although this control is not very rigidly exercised. Almost all its personnel are army officers and enlisted men on active duty. As a part of the army, the museum may call on other army offices for assistance, but by the nature of the museum's function there are natural limitations on the kind of useful services which can be provided. By regulation, all army regiments must send their flags and standards to the museum in case these units are terminated at any time, or if their colors are replaced with new ones. The army also will provide obsolete equipment, and arms manufacturers are required to send the museum one model of each new weapon manufactured, provided the museum can use it.

The museum operates essentially as a military organization. It has no board of directors or other policy-making body apart from its own staff. Naturally, any major policy decisions are subject to reversal by higher authority, but this seldom occurs. The director of the museum is its commanding officer in the full military sense of the term. He and the field grade officers assigned to the museum constitute the professional staff of the institution, and formulate general policy, although this responsibility appears to be exercised largely by the director and his assistant. The area in which the director can make decisions is indeed broad, for in most matters related to management of the museum he has full autonomy.

The functions of the museum are performed with a considerable degree of administrative simplicity. There are no formal departments providing a division of labor in accordance with some plan of specialization. In his capacity as general supervisor of museum activities, the director assigns the staff to particular sections where they presumably work until the expiration of their tour of duty. Certain functions also are assigned to the assistant director apart from the normal responsibilities inherent in his position. These are primarily related to detailed supervision of special museum projects. The museum's fiscal and clerical matters are handled by a special administrative section attached to the director's office and headed by a lieutenant colonel.

The substantive work of the institution is supervised rather uniquely. Each exhibit floor is regarded as a separate administrative entity with a field grade officer in the rank of lieutenant colonel or major in charge. Several company grade officers or noncommissioned officers serve under the floor commander, and each supervises a hall or one or two rooms. To carry the chain of command one step further, each hall and room supervisor has an enlisted man, normally a private, assigned to him to perform much of the routine work connected with the exhibits. If the personnel do not rotate in their museum assignments, there is some reason to believe they can build up a degree of specialization in particular categories of weapons or phases of Spanish military history. The workshop, laboratories, and library are supervised by the director's office and perform their services wherever needed throughout the museum.

The Museo del Ejercito normally has a complement of 50-55 persons assigned to its staff. Of the total, over 40 are on active duty with the army and about evenly divided between officers and enlisted men. The tour of duty the military personnel have at the museum apparently has no set length. It appears to vary with the individual, but rotation to other army assignments does occur after an average period of four years. Such rotation does present the museum with some difficulty in maintaining continuity in its staff. Assignment to work in the museum is considered choice duty, and selection of the individual is based on merit. No special talent for museum-related work, knowledge of arms, or demonstrated interest in Spanish military history is required for consideration.

Currently, the director is a lieutenant general and the assistant director is a colonel. Three lieutenant colonels and two majors are the other field grade officers. With the director and assistant director they comprise the museum's professional staff. By the nature of their duties these officers are similar to museum curators, but they do not necessarily possess the professional qualifications required for this rank. The remaining military personnel include 12 captains, 2 first lieutenants, 6 second lieutenants, 5 sergeants, and 14 privates. The rest of the employees are civilians and provide the museum with its secretary and technical staff. The six technicians include two master armorers, a joiner or woodworker, a phototechnician, a numismatist, and a master restorer of paintings. The armorers also instruct the soldiers in the use and proper care of the weapons in the museum's collections. Any work of preservation and repair required for uniforms is done outside the museum. The librarian is an elderly, retired army officer who volunteers his services to the museum and is thus not considered a member of the regular staff. The janitorial service is provided by six cleaning women. No special guards are employed at the museum. The soldiers on duty in the various halls perform this work. The military character of the institution is emphasized by one additional feature. The museum has an officer and non-commissioned officer of the day who remain on the premises for a full 24-hour period and comprise the night guard force.

Career museologists may question the advisability of manning a museum with military personnel who are not museum specialists and who are assigned to this duty for a comparatively brief period in their careers. This arrangement appears to cause no major disservice to the Museo del Ejercito, considering the stress given to the museum's custodial function. However, certain limitations in what the museum is able to do result from this personnel policy. These are felt primarily in the areas of research and services offered to the public, as well as the capacity to introduce modern museum practices. Since the museum does not have an exhibit specialist, any contemplated major changes in the displays would largely have to be wrought by outside assistance.

Because the museum seldom has specialists in weapons, colors, and the other fields of knowledge dealt with in a military museum, the Museo del Ejercito cannot undertake any program of systematic research. The staff does prepare an occasional magazine article and a monograph requested by the Military Historical Service and rather recently readied for publication a pamphlet on edge weapons. However, there is no established pattern to such research activities. The museum has recently completed its most ambitious publication project, that of preparing a 5-volume catalogue. These volumes are exceedingly well done and are perhaps the most impressive publication of this nature produced by a European military museum in recent years. Their completion was possible because of the concentrated attention the museum gives to documentation for items in its inventory. The staff also performs some research in answering questions sent to the museum. These inquiries are handled very informally and number about 300 a year.

Little assistance is given to the amateur collector of military objects. Weapons can generally be identified and historically described by the staff, but specialists must be called in if information is sought about uniforms and military pictures. No further professional service can be given. The museum does, however, help a small army museum in the Canary Islands and lends portions of its collections to exhibitions in Spain which commemorate military or historical events. It will not lend any objects to the public, nor does it provide guided tours for other than groups of soldiers and school children.

Financial support for the Museo del Ejercito is almost entirely provided by the Spanish Government. The routine operating expenses are budgeted for through the army, but the museum is not permitted to spend all the money it receives. It must divert at least 20 percent of its receipts into a special fund which accumulates to a sizable amount. It may then be drawn upon to pay for such major expenses as building repairs. Only the salaries of civilians employed at the museum are paid from the operating funds. The army pays the salaries of assigned military personnel. The museum charges a nominal fee for admission, but the income realized is comparatively small, since only about one-third of the visitors must pay to enter. The money received is placed in the operating funds and may be used to help meet museum expenses. The catalogue and guide book are sold at the sales desk, but no income is realized from this venture.

The Museo del Ejercito is of considerable interest to the public, for more than 190,000 persons visit it each year. It is conveniently located in the heart of Madrid and near three of the city's major tourist attractions—the gardens of the Retiro Park, the old church of Jeronimo, and the Prado, one of the most celebrated art galleries in Europe.

MUSEO NAVAL
Madrid, Spain

The first allusion to Spanish naval history normally prompts one to recall immediately an age which is now almost four centuries past. This was the period when the Invincible Armada brought Spanish naval power to its zenith, when Spanish vessels carried explorers to the New World and other far-flung outposts and soon brought troops to stabilize the rule of the Spanish crown in its new colonial possessions. After the English defeated the Armada in 1588, Spanish naval power began to decline steadily, even as Spain began to diminish as a military power of consequence on the European continent. This decline was temporarily halted early in the 18th century when Spain fashioned powerful naval convoys to guard the shipment home of newly discovered treasures in Mexican silver. However, by the early 19th century, Spain could not halt the burgeoning nationalist movements in Latin America or the surge south and west of the American pioneer to the north and relinquished all but a few minor holdings in the Western Hemisphere. At the end of the 19th century, most of these, as well as its Pacific possessions, were lost when the United States defeated Spanish fleets at Manila and Santiago during the Spanish American War. The Museo Naval in Madrid concentrates its attention upon the age of Spain's major naval achievements and thereby creates the general impression that it displays little more than a collection of antiquities.

Judged from the contents of the museum, many generations ago the Spanish Navy saw great value in collecting and preserving significant mementos, relics, and documents. It is difficult to discover whether these materials were first assembled by official direction or by interested naval officers who bestowed their collections upon the navy, thus placing it in the museum business. The date when the navy began to let the public view its "cabinet of curiosities" is likewise somewhat obscure, although the Museo Naval probably has been in existence, in one form or another, as long as its army counterpart, the Museo del Ejercito. It has been in its present location, the Ministry of the Navy, since 1929, the date the building was completed. Prior to that it occupied rooms in the Palace of Peace, which has since been torn down to make way for a new thoroughfare.

A visit to the Museo Naval raises some question as to whether its inclusion in the Navy Ministry building was originally contemplated. If at some future date the Museo Naval might well be placed in its own building, its present rooms could be converted to office space with a few simple alterations. The exhibit space does not appear to be permanent, for individual rooms are formed by drapery dividers. Modern museum techniques would also be difficult to employ. In short, the arrangement and appearance of the Museo Naval readily lead to the conclusion that an office building is not an acceptable place to house a first-rate museum.

The museum exhibit halls, library, and offices occupy approximately one-half of the ministry building's main floor. There is an auditorium on the second floor, and the workshop and laboratory are elsewhere in the building. The exhibit space is compact and features a central hall, surrounded by a number of small rooms. Most of these are accessible from the central hall, but they are also interconnected to facilitate the visitor's tour through the displays. The main room occupies a fairly large area, and its ceiling is actually the roof of the building some three stories above. The museum rooms opening onto it are only one story in height. Their separation from the central hall is more apparent than real, and is accomplished by widely spaced square columns. Because of the ceiling height in the central room, the museum is able to exhibit several large ship models with all sails and rigging, as well as a number of flags and colors.

A modified chronological plan of presentation is followed in the exhibits with major emphasis given to the Spanish Navy's most prosperous periods. In the first rooms relics of the early voyages of discovery are preserved, and these are followed with references to naval victories in the 16th century. Defeats dealt the navy understandably either are lightly glossed over or ignored, and so a comprehensive presentation of the sweep of Spanish naval history is lacking. References to the navy of the present or the recent past are at a minimum. Interspersed with the chronological halls are several topical halls which display various insignia and medals, arms, and projectiles. The halls are clearly numbered, and so if the visitor adheres to the plan suggested for his tour he will be able to follow the chronological sequence intended.

Since the Museo Naval has not been allotted a great deal of exhibition space, its rooms are quite small and most are fairly crowded with objects. There is ample space for visitors to move about, but a very large crowd cannot be accommodated too easily. Most cases contain a sizable number of artifacts, but they are not cluttered. Many items are of great historical interest. The confined space of the exhibits provides the museum with little flexibility. Hence, the displays are seldom changed or relocated, although an occasional accession is added to its pertinent exhibit.

By comparison with some other European naval museums, the inventory of the Museo Naval is not large. The museum records reveal a collection consisting of 50 uniforms, 200 small arms, 60 edge weapons, 20 cannon, one piece of mobile artillery, five torpedoes, two mines, about 100 projectiles of various types (mostly ancient), and approximately 100 medals and insignia. Its total military model collection consists of approximately 300 ships models and four aircraft models. Completing its inventory of artifacts are two collections of personal arms belonging to famous Spanish naval commanders. These are catalogued separately. Other donations have been combined with the museum's general collection.

One of the museum's most valuable possessions is its library, for it has gathered many items of historical significance which are invaluable aids to research. The library contains a rather extensive collection of original manuscripts belonging to the early Spanish explorers, as well as the greatest known collection of 16th-century nautical books. The nautical books are especially valuable, for they describe the art of navigation as it was practiced during the time of the great explorations.

The exhibits of the Museo Naval occupy approximately 21,000 square feet of floor space, an area slightly larger than that used by the Sjohistoriska Museum in Stockholm. Because the museum is housed in an office building, it is not permitted the luxury of any exterior exhibit space. The amount allocated for storage is not known, although it is believed to be negligible. The library is moderately sized, occupying about 1,100 square feet, and a study room of 500 square feet provides the staff and other students fairly adequate working space. The museum likewise is fortunate in having reasonably large offices which occupy about 2,300 square feet. The laboratory and workshop are fairly small and total only 700 square feet. Other service rooms occupy a space of 700 square feet, and a sizable auditorium of 2,700 square feet gives the museum an opportunity to conduct lectures and other special events.

The pattern of governmental control exercised over the operations of the Museo Naval is very similar to that employed by the Spanish Army in supervising the activities of the Museo del Ejercito. The tie-in to the patron service is very close for both museums, and in the case of the Museo Naval the link is strengthened by the physical co-location of the Navy Ministry with the museum itself. At least the museum is sufficiently close at hand should naval officials wish to keep an attentive eye on its operations. In the administrative structure of the navy, the museum is directly attached to the office of the chief of general staff of the navy within the ministry. As in the Museo del Ejercito, most of the personnel in the Museo Naval are officers or enlisted men on active duty. The museum may likewise call upon the navy for a reasonable amount of useful assistance, and by legislation the navy is required to send the museum all ships models or any other object of historical significance which it discards from use or obtains by donation.

Although the Museo Naval is staffed largely by naval officers and enlisted men, it is not strictly a military organization as is the Museo del Ejercito; in its operations it is more comparable to the Musée de la Marine in Paris. The major policies of the museum are determined by a 20-member board of directors. The board's ex-officio members include the chief of general staff of the navy, the commandant of the Madrid Naval District, the director of the Museo Naval, and the Minister of the Navy who serves as board chariman. The remaining members are prominent citizens who have demonstrated a keen interest in the museum and naval matters and descendants of naval heroes whose presence on the board gives it added prestige. The Minister of the Navy appoints all but the ex-officio members. It approves the budget and resolves all economic questions involving the museum in rather general terms. A majority of the museum's routine activities, of course, are handled by the director.

The Museo Naval enjoys much organizational flexibility. There are no formal departments separating the museum's substantive functions, and each naval officer assigned to the staff is permitted to pursue his own particular field of interest under the general guidance of the director. No apparent conflicts result, since there is considerable diversification of interest within the present staff. The library and workshops, including the exhibit laboratory, are regarded as separate entities under the immediate supervision of the director. The museum also has an administrator who functions as a business manager. It lacks an exhibits division, but the officers make any changes required in those displays which are pertinent to their field of individual activity. As in the Museo del Ejercito, routine work on the exhibits is performed by the enlisted men, with each sailor responsible for a particular room.

The naval officers assigned to the museum can remain for an indefinite period if they meet the standards of performance required and if they desire to pursue the career of a naval museologist. A decision to remain at the museum may work to the officer's disadvantage insofar as his opportunity to rise in rank is concerned. Promotions in grade and responsibility are made only when vacancies occur at the top. Since the group of officers at the museum is very small, the turnover is slight, and a younger officer may remain at the same military rank for a considerable time. The chief advantages derived from a prolonged personnel assignment are, of course, the opportunity given the individual to develop a field of museum expertise, the strength added to the museum's research program, and the assurance of staff continuity.

Thirty-five persons are presently on the staff. Five of these are officers with duties comparable to those of a museum curator. The director, who holds the permanent rank of naval captain, also serves as the director of the Instituto Histórico de Marina (Institute of Naval History) and superintends the naval archives at Santa Cruz. The latter responsibility requires little effort, and the archives are managed as a museum branch office. The museum has an assistant director, and the other three officers respectively are specialists in cartography, ship models, and astronomy. Two librarians, both civilians, supervise the museum's extensive collection of manuscripts, documents, and other research materials. The four clerical employees and seven technicians are also civilians. The technicians include three carpenters, a painter, a draughtsman, a restorer of paintings, and a photographer. Thirteen sailors are assigned to the museum to care for the exhibits, but their duties go little beyond keeping the displays clean. They also provide the museum with its guards. The only other employees are two charwomen.

Because the Museo Naval maintains a close relationship to the Institute of Naval History and also possesses an excellent library, it functions as an active research center. All the officers conduct investigations directly related to their museum work or in some specialized area of naval history. Some of the research projects are extensive and can be undertaken without fear of transfer to another assignment before the work has been completed. The professional staff of the Museo Naval has written an impressive number of books as well as articles which have appeared in the principal navy magazines and other professional journals. The chief areas of inquiry have been geography, astronomy, naval construction, and the history of navigation, voyages, and discoveries. The museum has performed a particularly valuable service for scholars by systematically compiling documents of the early Spanish explorers. Recognizing that the size of its staff limits the amount of needed research which can be undertaken, the museum encourages other interested students to use its facilities. A considerable number constantly avail themselves of this opportunity.

The number of services the Museo Naval offers the public is somewhat greater than might be anticipated for an institution of its size and scope. The staff conscientiously attempts to answer questions submitted by the public and presently handles about a thousand a year. The museum likewise extends its assistance, whenever possible, to the amateur collector of naval objects. The staff identifies items brought to the museum, but it can offer little help in restoration or preservation. The Museo Naval gives one unique type of assistance to the amateur collector which is not provided by any other major military museum in western Europe. It occasionally sets aside a portion of its very limited exhibit space so that private collectors can display some of their most prized objects.

In common with most other museums, the Museo Naval will lend some objects to other museums or for special exhibitions where naval items are requested. Students may use but cannot remove reference materials from the library, but occasionally books are lent to naval officers. Because the Ministry of the Navy has provided it with an auditorium, the Museo Naval has been able to conduct public lectures and show motion pictures. In the past, this program has been sufficiently elaborate to include the presentation of full lecture series. The museum has also sponsored an occasional conference on naval history. One further service offered to the public is a guided tour of the exhibits. However, the tours must be scheduled in advance and are conducted only for groups or special guests.

Since the museum is an agency within the Ministry of the Navy, its financial needs are budgeted for through the ministry. Actually, the museum has two budgets. One provides for routine expenses and covers such items as care of the exhibits, purchase of supplies, salaries of civilian employees, and the acquisition of books, pictures, models, and other objects. The other budget is provided to deal with extraordinary expenses and is established at the request of the museum when the need arises. It likewise varies in amount and is submitted only when building repairs are required or when the museum has insufficient funds to pay for unexpected, but necessary purchases. The salaries for the museum's naval personnel are paid directly by the navy and are not incorporated in the museum's routine budget. There is a nominal charge for admission which nets a small income. Part of the money received is placed in a reserve fund and applied against museum expenses, and part is given to the lesser paid employees to augment their incomes. The only other source of recurring revenue is from the sale of prints, pictures, and some publications.

The Museo Naval is centrally located in what may be regarded as the heart of the Madrid tourist area and attracts an average of 30,000 visitors each year.

MUSEU MILITAR
Lisbon, Portugal

The military and naval history of Portugal closely parallels that of its neighbor Spain. Both countries fought the Moors for generations, but the Portuguese succeeded in driving their common enemy back to the natural frontiers of Portugal some years before the Spanish won their final victory. Thus, Portugal was able to divert its attention elsewhere at an earlier time, and under the able leadership of Prince Henry began the organized exploration of the west coast of Africa. By the time Columbus undertook his voyages in the name of Ferdinand and Isabella, Spain's armed might was at its peak, whereas Portuguese military strength had declined as the nation concentrated its efforts on becoming a great maritime power.

A potentially explosive situation in the growing rivalry of the two nations was averted in 1493 when Pope Alexander VI proclaimed a line of demarcation which granted Spain the exclusive right to possess all lands west of a line drawn 100 leagues to the west of the Azores. As stipulated in the Pope's decree, Spain's possessory rights extended as far west as India. Portugal protested the location of the line, if not the principle of deciding such matters by papal fiat. The two nations then agreed to move the line 270 leagues farther west and solemnized this decision in the Treaty of Tordesillas in 1494. The practical effect of the agreement was to give Portugal possession of Brazil and to send Portuguese explorers and soldiers to carve out colonial posesssions in Africa and India. Portugal managed to build a flourishing East Indian trade which existed virtually as a monopoly until the Dutch began to make heavy inroads a century and a half after Vasco da Gama touched the shores of India.

The record of Portuguese military achievements has been preserved at the very interesting Museu Militar located at the edge of the Lisbon harbor on a busy square now called Largo do Museu d'Artilharia. A separate navy museum is presently being organized in Lisbon, but it is not yet open to the public. The similarity in content and philosophy between the Museu Militar and the Museo del Ejercito in Madrid is inescapable. To a lesser extent the comparison also holds for the Museo Naval. The major emphasis is quite naturally placed upon Portugal's age of greatest glory and is rather well symbolized by one of the first objects seen when entering the Museu Militar. This is the sword of Vasco da Gama, the one which the intrepid explorer is alleged to have carried with him as he led a Portuguese expedition to India. The flavor of the past lingers in a great many of the halls, although this museum has attempted to present all the significant epics of Portuguese military history from the late 15th century to World War I.

The Museu Militar first began as a collection of artillery, but its scope was expanded in later years to include other facets of army activity. Still later an effort was made to place the exhibits in some historical context. Some chronology has been observed in planning the displays, but for the most part the halls do not follow a precise time sequence. Thus, the viewer cannot see the unfolding of Portuguese military history as he moves from hall to hall. Rather he observes a mixture of historical references, offered amid an abundance of arms and armor.

The building housing the museum appears to be in the style of early rococo architecture popular in the late 17th century. Its exterior is somewhat plain, except for the front and side entrances. These are topped with ornate military statuary and Portuguese coats of arms. The structure is rectangular with a sizable open interior court. The rooms are all interconnected in the familiar style of 17th-century palaces. Wall space is something of a premium, since both the interior and exterior walls in most rooms are lined with closely spaced floor to ceiling windows. As a result, most of the museum's paintings are hung at either end of the rooms or between the windows where space permits. Many halls are extremely ornate, and their rather florid ornamentation is further accentuated by the method of display employed. Virtually all the ceilings are decorated in a late baroque or early rococo style. Many of the rooms have concave transitional lunettes between the walls and ceilings. These and the ceilings are frescoed, carved, or painted or contain elaborate bas-reliefs. The over-all impression is one of opulence and past grandeur.

Some exhibits of the Museu Militar show little evidence that most recent museum techniques have been applied. Perhaps the building design, with its paucity of wall space, has made it impossible to prevent the evident overcrowding found in many halls. However, the superabundance of objects displayed in rooms of moderate size yields the natural impression that the museum keeps as few items in storage as possible. All exhibits are very compact. Only small showcases are used, and these are reserved for significant and valuable specimens. Many objects are placed in the open on tables, where they can be closely examined. The museum's favorite method for displaying rifles is to rack them along the wall, one tier above the other, or to stack them on the floor in groups of three. Where this is done the exhibit room resembles an arsenal more than a museum. A room whose walls are lined with arms and whose ceiling is magnificently decorated in the baroque style presents a somewhat incongruous sight. Individual pieces can only be examined with major difficulty under such conditions, and the tendency on the part of some visitors may be to pass swiftly on to the next room without making a real effort.

An excess of quantity exists in all parts of the museum. Additional examples of this are the number of spears, pikes, swords, and other edge weapons grouped high up on the walls in interesting and quite attractive artistic designs. This form of display may be esthetically pleasing, but it seems somewhat inappropriate to a museum which tries to give its visitors a better understanding of Portuguese military history and an opportunity for a close examination of arms and armor. One other unique method for displaying military trophies is worthy of mention. Crossed swords, battle axes, and rifles have been used as legs for the small showcases or as braces for the legs of larger cases. Objects used in this fashion have no labels, and, indeed, many throughout the museum are without labels.

The halls of the Museu Militar are named rather than numbered. Some are designated for former rulers, such as Jose I, Maria II, and Prince Henry. Still other halls commemorate famous commanders and explorers such as Alphonse de Albuquerque and Vasco da Gama. The names of another group of rooms recall major military engagements or areas of Portuguese military achievements. Examples of these are the hall of the Peninsular War, the hall of the Campaign for Liberation, the Great War (World War I), the European hall, and the African hall. In the commemorative halls there is a splendid use of art work, and some of the paintings and murals are most impressive.

Throughout the Museu Militar the preservation of arms and armor compares very favorably with that found in other European military museums. Some of the flags and colors are in a good state of repair, although they are not always displayed to full advantage. Many are displayed vertically and in groups, and so it is quite impossible to see their designs. They likewise are not encased, and are thus subject to the continued deterioration caused by constant exposure to the air. The general upkeep of the exhibits and building appears fairly adequate, although it is evident that significant changes seldom occur. The exhibits are all considered permanent and are modified only with the addition of an occasional new acquisition. Consideration has been given to the need for additional exhibits in the future, for three rooms have been reserved for displays depicting subsequent events in Portuguese military history.

The museum's great emphasis upon collecting artillery and small arms at the time of its origin is evident not only upon examination of its exhibits but also from a listing of its inventory. Its collection of small arms now contains over 5,000 individual items. All but a few hundred are on display. The museum possesses 620 swords and other edge weapons, whereas its heavy ordnance collection contains 257 cannon and 262 pieces of mobile artillery. Other items in the inventory include 155 uniforms, 525 insignia and medals, 70 military models of various types, a sizable number of musical instruments, a moderate quantity of flags and colors, many paintings, and a modest library. The museum has received some personal souvenirs and mementos of famous Portuguese military leaders, but these have been merged with the rest of the objects and are not maintained as separate entities.

The Museu Militar contains a fair amount of exhibit space, for the collections are housed in 33 rooms of varying sizes, though none is particularly large. It is estimated that these halls provide 25,000 to 30,000 square feet for inside display. The interior open courtyard has an area of 10,000 square feet and functions partially as an artillery park, although not too much of the space is now being used. The staff maintains no separate reference collections for special study, but will work either with items already on exhibit or with individual objects which are not of sufficient size to warrant special storage arrangements.

Even though this museum appears to have placed most of its inventory on display, it still allocates a considerable amount of space for storage. In storage capacity it is much better off than most military museums, for it is able to allocate six rooms in the building for this purpose plus a nearby warehouse suitable for storing particularly large items. It is likewise well equipped with service rooms. The museum has five private offices for staff use, two sales rooms, six employees' rooms, an employees' lounge, two security vaults, a workshop, and a library. It has no auditorium or study rooms other than the small library and the private offices of the staff.

The Museu Militar is under the jurisdiction of the Portuguese Army and administratively is attached to the general staff. There is little evidence to indicate that the army gives very close scrutiny to museum operations. The pattern of control it exercises appears to be more in keeping with that found elsewhere in Europe— general fiscal supervision with routine operations left to the discretion of the director and his staff. In addition to a general review of over-all museum policies, the army contributes objects to the collections of the Museu Militar and renders other needed services which the museum is unable to provide from its own resources.

The director and his immediate staff set the major policies of the museum since there is no separate board appointed for this purpose. The director apparently is vested with considerable personal autonomy in governing museum activities, but he is assisted in his tasks by an administrative council, over which he presides and whose members include the subdirector and several other assistants known as adjutants. These are equivalent in rank to museum curators. The administrative council meets daily. The professional employees of the Museu Militar, all of whom serve on the council, are selected from among the ranks of reserve artillery officers. To qualify for selection they must have revealed some competence in the field of military history, art, or related cultural pursuits. Prior museological experience is not a requirement for consideration, however. Considering the size of the undertaking, the museum has relatively few employees apart from the professional staff. Five technicians of varying skills attend to the care of the collections and the upkeep of the exhibits, whereas the administrative and clerical work is handled by three individuals. Two janitors are the museum's only other full-time employees.

Because the routine activities of the Museu Militar are rather closely supervised by the administrative council, the need for an extensive formalized organizational structure has largely been eliminated. No special departments have been created to handle arms and armor, ordnance, uniforms, colors, or other specialized areas of work germane to military museums. Instead, the office of the director, the administrative council, a section entitled general services, and the library are listed as the museum's principal administrative elements. The general services section embraces the work of the technicians on the staff, and though the library is carried as a separate organizational unit, the museum employs no librarian. Presumably, it is administered by the director's office or one of the adjutants.

The program of the Museu Militar does not appear to be as dynamic as those existing in similar institutions elsewhere, nor does the public support its activities too extensively. The museum is located in the heart of one of Lisbon's major tourist areas and draws on the average 14,000 visitors a year. It is possible that the limited nature of its collections restricts its appeal largely to individuals who are seriously interested in military history or to those who enjoy looking at military objects.

The remaining services provided by the museum are similar to those performed by other military museums. Research is restricted to the study of those specimens now in the inventory of the Museu Militar which have not yet been identified. The information gleaned from this type of research is primarily retained in the form of data for the museum's catalogue and indices. The library at the museum is restricted to staff use only. However, assistance is given to private collectors in the way of identifying specimens or providing other information needed by the amateur. The Museu Militar also makes its services available to the local museums which army units operate. Its principal assistance is rendered to those organizations through the lending of objects needed for special exhibitions or for study.

As a government agency the Museu Militar depends almost entirely upon the public treasury for its operating funds. These are budgeted for by the army and included as part of the annual appropriation request. Fifty percent of the museum's outlay is allocated for salaries, 35 percent for new acquisitions, and the remaining 15 percent is applied to the cost of maintenance and supplies. The museum has two other sources of revenue, both of which are fairly minor. A nominal fee is charged for admission, and the income received is added to the general funds of the museum. Some profit is also realized from the sale of publications and post cards. This too is applied to the museum's general operating funds.

SIGNIFICANT SPECIAL ARMS COLLECTIONS

Although almost every European country has a major national military museum, there are many interesting and significant weapons collections housed in smaller and local museums. Some have been given a place of prominence within other outstanding institutions whose range of presentation includes several fields of knowledge. A number of these originated as the personal collection of some military hero or as the mementoes of an army unit. Still others were formed from the personal arms and armor of kings and emperors. These often have been preserved as a single collection and displayed in or near the royal palace. Many city and provincial museums have also sprung up throughout Europe. Virtually all tend to record the significant events of local history. Since many of these cities and provinces were once autonomous political units, they produced some very absorbing chapters of military history. These have been duly subjected to some form of graphical presentation.

The unit, regimental, or divisional museum has been popular in Europe for a number of years and is steadily growing in popularity in the United States. For example, British regimental museums are particularly excellent. By charting the regiment's achievements in battle, these museums contribute much toward preserving venerated traditions. Their collections are personalized because members of the regiment actually used the equipment displayed during their years of service. Such a degree of personal identity is most difficult to achieve in the larger national museums. Elsewhere in Europe the unit museum is most often found at army installations, although some small navy museums are occasionally located at major ports.

The type relationship maintained between the unit museum and its national counterpart varies from country to country, but it is usually quite informal. The local museum is seldom self sufficient with respect to highly technical matters. Hence, it will often seek assistance from the national institution, particularly in restoring and preserving specimens. Frequently, the national museum serves as a source of expert consultation in other matters and generally lends objects for special exhibits or for study. The objectives of the regimental and unit museums are, of course, often rather limited. Such museums seek primarily to preserve objects of only local significance. Their preoccupation with military history seldom extends beyond describing the unit's achievements in battle. Seldom do they seek extensive public patronage. Because unit museums are so specialized and often quite inaccessible to the public, they are spared many of the problems with which the national museum, the military division of a general museum, or other specialized collections of national significance must cope.

In those countries where there is no national military museum, the museological presentation of military history is diffused among several local institutions. Usually the subject is given some treatment in special galleries of general museums. As a result, the focus of the exhibit tends to be provincial, and the approach to military history is often regional. This form of exhibition together with the specialized weapons collection invites comparison with the national military museum. They share much in common, but they do have marked differences.

As in the case of the national military museum, each military department within a general museum must develop its own philosophy. However, the fundamental decision on such matters is sometimes out of the military curator's hands, for as a subordinate official he must fashion his displays within the context of the over-all museum theme. Since many general museums tend to provide a strong historical emphasis, the military galleries reflect a similar concern. Hence, the viewer is very often treated to something more than a mere display of arms and armor—he also learns some military history. The custodial function is not depreciated, to be sure, but those objects which are historically significant and which underscore the museum's over-all theme must be selected for display.

The problem of determining a philosophy is considerably reduced for the curator who presides over a specialized collection. If the personal mementos of a celebrated commander or monarch are to be exhibited, the curator's responsibility is rather clear—he must prepare a display which not only shows off the collection to its best possible advantage but also suitably commemorates the military achievements of its famous owner.

Regardless of the type of museum in which they work, military curators must confront other identical problems. They must always determine what specifically to show, what to retain for ready reference, and what to place in reasonably permanent storage. Simultaneously, they must seek to utilize most effectively the space available to them. This can be a serious problem to the military curator in a general museum, who must compete with his colleagues for valuable space which always seems insufficient to meet everyone's needs.

The amount of research which can be profitably undertaken is another common problem and is related to the all-important factors of time, number and competence of the staff, and adequacy of facilities. The curators in many smaller museums closely parallel their colleagues at the national military museums both in the quantity and quality of their scholarly achievements. Some enjoy an international reputation and have produced a number of well-received articles and books. All generally make their services and fund of information available to amateur collectors and other students of the weapons and panoply of warfare.

A final problem which every military curator confronts is the restoration and preservation of military objects. The larger national military museums are often adequately equipped to meet their basic requirements, and can restore all types of weapons, colors, and paintings. The curator in a general museum may be able to meet some of his needs from the resources of his museum, but not often can he do everything within his own workshop.

Thus, he may have to draw occasionally upon outside agencies for assistance.

The departmental status of some military collections imposes certain practices and some administrative arrangements which differ to a degree from those of a national military museum. First, there is a strong tendency to specialize. The collection itself may force some specialization, but more often space and personnel limitations cause a certain amount of concentrated effort in a particular category or subject area. With only a few halls at his disposal (sometimes not more than two or three), the military curator must carefully evaluate his collection and choose what he believes is significant. He seldom has sufficient space to display heavy ordnance. Hence, small arms, some armor, uniforms, and heraldry usually comprise his exhibit. If the curator is a specialist by training, he can hardly resist displaying those items with which he is most familiar. Further, where the curator must exhibit his collection within the confines of a provincial historical museum, he can seldom demonstrate the evolution of weapons, describe tactics and strategy, or expand his frame of reference beyond the local area in any substantive field of inquiry he might like to explore.

Most national military museums are comparatively well staffed with experts in the required fields of knowledge. A group of technicians who can provide almost every service needed augment these professionals. This desirable situation, of course, cannot exist in the general museum. Usually only one military curator is on the staff, although he frequently has an assistant. He very often is a specialist in arms, armor, or some other field, but he has to perform the functions of a generalist. He likewise has a small number of technicians who work with him. Thus, when he cannot meet his needs with the resources at his disposal, he must obtain professional consultation elsewhere or improvise to the best of his ability.

The military curator in a general museum cannot experience the autonomy enjoyed by the director of a national military museum, for he is at the administrative level of a department or division head and is fully subject to the managerial policies of his institution. Yet, in the daily operation of his department, he is naturally allowed considerable freedom of action. He likewise is relieved of many administrative problems confronting the head of his own establishment or the director of a major military museum. For example, he normally depends upon the administrative services performed by the museum for its employees, rather than having to provide them for himself. Further, his budgetary responsibility is restricted to his own personal concerns and is exercised within the competitive atmosphere of an institution with several departments. His request for operating funds is evaluated against similar requests from his other colleagues on the staff, and the attitudes of the director and museum board toward his operations are all important in determining what share of the over-all museum budget he will be allocated.

There are many general museums with military divisions and special military collections which underscore these observations. A few of the more important ones will serve, however. They are divided into three categories. The first group of five consists of military collections housed within a general museum. They are: the Museum für Hamburgische Geschichte in Hamburg, the Bayerisches Nationalmuseum in Munich, the Schweizer-Landesmuseum in Zurich, the Historisches Museum in Bern, and the Musée d'Art et d'Histoire in Geneva. The second category is closely related to the first, with the exception that the military collections are highly specialized. They likewise are exhibited within a museum which treats a variety of subjects. The two collections discussed are Die Waffensammlung (Hapsburg armor) located in the Kunsthistorisches Museum in Vienna and the Oldsaksamling (Viking finds) in the University Museum in Oslo. Special collections housed as separate museums, but which do not have the magnitude or scope of a national military museum provide the third category. The collections located at the Castel San Angelo in Rome and the Real Armería in Madrid furnish useful illustrations.

MUSEUM FÜR HAMBURGISCHE GESCHICHTE
Hamburg, Germany

Prior to World War II the Berlin Zeughaus provided Germany with a fairly representative national military museum, for its collections covered the span of Prussian and modern German military history. When the war ended and Germany was divided politically between East and West, such a national military museum ceased to exist. At least this is true in West Germany. When the Zeughaus reopened under its new title, the "Museum für Deutsche Geschichte," an institution possessing some of the attributes of a national military museum was created for East Germany. As presently constituted, this museum fulfils an educative function in behalf of Marxian ideology, and for this reason it offers a Marxist version of German military history which West Germans cannot accept as either accurate or objective.

The evolution of pre-World War II Germany from a group of independent and semi-independent political divisions is reflected in the type of military museums which have developed in what is now West Germany. Allegiance and devotion to the history and culture of the local area, city, or province have stimulated the rise of several fine museums which seek to commemorate past achievements and venerate local traditions. Since much local history involved military activity, these museums have their military galleries. The German people have long been ardent collectors of military materials, and the collections displayed in the various museums throughout the country not only are excellent but reflect the use of modern museological techniques.

One of the best examples of a city museum is the Museum für Hamburgische Geschichte (Museum for Hamburg History). Hamburg at one time had the political attributes of a city state, and during the course of its long history has been a major center of trade. The power of the city was most effectively symbolized by its great commercial achievements and its shipping companies. To protect its political status, the city also developed some militia and naval units. Thus, the history of Hamburg contains a number of interesting military chapters. However, the city never acquired a reputation for great military prowess.

The Museum für Hamburgische Geschichte is one of the largest and best regarded city historical museums in western Europe. Although the building was constructed several decades ago, it is most responsive to the application of modern museum techniques and is both impressive and attractive. Many of the halls are spacious, well lighted, neat, and orderly. Excellent use is made of wall space, and no room contains more than a few cases. Individual showcases are not cluttered with objects, and

Museo del Ejercito, Madrid. The hall displaying ordnance models.

Museo del Ejercito, Madrid. One special feature of the Museo del Ejercito is its hall commemorating the contributions women have made to the nation's military history. Most of the women memorialized were heroines of the Franco forces during the Civil War of the 1930's. As shown here, the hall is essentially a portrait gallery, although a few personal mementos are also displayed.

Bayerisches Nationalmuseum, Munich. The open Artillery Court featuring the finest cannon in the Museum's collection.

Bayerisches Nationalmuseum, Munich. The hall displaying flags and colors.

significant and precious items are displayed in the manner of a fine-arts collection. All ship models are individually cased, and only representative ones are shown. The museum makes rather extensive use of models and dioramas to illustrate the city's commercial and transportation activity but does not employ such techniques in its military halls.

An evaluation of the museum's exhibits readily produces the conclusion that there is a sincere effort to provide the viewer with a representative picture of Hamburg's history. For this reason, comparatively little space has been given to the display of military objects. Nevertheless, the coverage of Hamburg's military history seems rather comprehensive. The collection basically consists of some uniforms, small arms, edge weapons, ceremonial weapons, a few artillery tubes, representative flags and colors, and personal objects used by members of the local constabulary. The exhibits follow the same general pattern of those elsewhere in the museum, and all objects are in excellent repair.

The atmosphere at this museum is strongly academic. The nature of the institution, as well as the German museum tradition, insures this great preoccupation with scholarship. The director and his principal assistants are well-trained historians and hold the Ph.D. degree. The director is also a professor at the local university, and other members of the staff lecture there. Research is one of the major activities at the museum, and the military curator devotes considerable time both to research on local military history and to the study of weapons. His duties also extend to other museum matters, for care of the military collection is not a sufficiently large responsibility to require full time attention. Hence, the military galleries are kept within their proper perspective in relation to the museum's other departments, but they are maintained in accordance with a very high standard of professional excellence.

BAYERISCHES NATIONALMUSEUM
Munich, Germany

The Bayerisches Nationalmuseum (Bavarian National Museum) is one of the very best German provincial or state museums. Located in Munich, the capital city of the old kingdom of Bavaria, in southern Germany, it draws upon the rich history of the region for the subject matter of its exhibits. The military history of Bavaria is extensive, for it was one of the dominant German electorates in the Holy Roman Empire and a military factor of some importance in central Europe for many generations. Since World War II the Bayerisches Nationalmuseum has allocated several halls to display a small portion of the very considerable collection of weapons and armor which have been assembled from the past.

Prior to World War II Munich had its own large impressive military museum, and the classical dome-crested building housing its ample collections provided the city with one of its leading tourist attractions. During the war the building was badly bombed. Although much of Munich has now been rebuilt, the museum's ruins still stand as a silent reminder of war's destructive power. At the beginning of hostilities the staff stored the collections in the museum basement or in other safe places. Hence, the original inventory remains pretty much intact. With the removal of West Germany's occupation status a few years ago, the way was opened once again for the display of military materials. Pending a final decision on the disposal of the old military museum building, space was set aside in the Bayerisches Nationalmuseum to display part of its collections. Since no plans for a separate military museum appear to be in the offing, the collections will probably remain in their present location for an indefinite period, with much additional exhibition space now contemplated.

The Bayerisches Nationalmuseum is exceedingly attractive and well kept. Architecturally it is a rather ponderous composite structure with some classical lines. Some of its rooms have a Gothic motif with arches and vaulted ceilings. These provide a natural setting to display suits of armor. At present there are only four halls devoted exclusively to the display of military objects, but plans are underway to obtain an additional 15 rooms within the next two to four years. This added space will be sufficient to exhibit the best pieces formerly shown in the old museum.

The present military galleries give chief emphasis to a display of 17th- and 18th-century arms and armor; virtually nothing is shown which is reasonably modern. Suits of armor are displayed

SIGNIFICANT SPECIAL ARMS COLLECTIONS

in a line along the walls. Nearby in racks are pikes, halberds, swords, and other associated weapons. In two of the rooms rows of helmets are displayed on the upper walls and out of the range of close observation. The armor is reasonably well identified, and the wars in which it was used have been duly noted, but beyond this the exhibit is devoid of a historical context. Perhaps more will be done with military history when other halls are made available.

The Bayerisches Nationalmuseum has an interior courtyard which is now used as a small artillery park. It contains two rows of 10 cannon mounted on concrete blocks. The cannon have been selected with great care, for they are the finest and apparently most ornate pieces in the ordnance collection. They likewise represent major developments in the evolution of artillery.

The museum has one additional military hall which displays flags and colors, a few medals, and some insignia. The method employed to exhibit military colors is unique but not very satisfactory. There is a ridgepole arrangement in the center of the room with a row of closely spaced flags mounted on staffs extending from either side. The staffs are tilted slightly upward rather than extended horizontal, causing the flags to have a few folds. The smaller standards and guidons are similarly mounted along the wall. Both are sufficiently high to permit a person of average height to walk beneath them. This form of display makes it virtually impossible to see an individual color without separating it from the one hanging at either side. Dissatisfaction has been expressed with this method of exhibition, but it does provide a simple solution to an acute storage problem and prevents any rapid deterioration of the collection.

The head military curator was a top official of the former arms museum. In some respects he has again had to start from the beginning. Where once he made policy decisions affecting the operations of an entire museum, he now is in the position of a department head. Because his collections are sizable, he undoubtedy finds it somewhat difficult to adjust to the condition of vastly reduced exhibit space. Within his new status, of course, he has been relieved of many managerial problems and permitted to plan new halls which will return a fine weapons collection to public view.

SWISS HISTORICAL MUSEUMS

Because there is no single national military museum in Switzerland, the exhibition of military collections in that country follows a pattern somewhat similar to that found in West Germany. The principal cities of Zurich, Bern, and Geneva each have a sizable museum, and included in each is a moderately sized weapons collection. All three are comparable in type of subject matter, although the Schweizer Landesmuseum in Zurich appears to have been built on a somewhat grander scale and gives much more attention to general Swiss history. Hence, its scope is considerably less provincial.

Also scattered throughout the country are other interesting military collections. The armory in Solothurne, the rather extensive collection of ancient and modern weapons found in the interesting old Schadau Castle in Thun, the arsenal at Morges, and the arms found at the Chateau de Chillon in Montreux are some of the more prominent ones. Virtually all these collections antedate the 19th century, for the Swiss people have been spared participation in war since their nation attained the status of international sanctuary.

The Schweizer Landesmuseum building gives the appearance of considerable antiquity, for it resembles a combination medieval castle and town hall. Actually it was built only several decades ago. This type of construction has produced some difficulties in the exhibit halls, at least in applying modern museum techniques, but it offers some esthetic advantages in exchange for some obvious inefficiency.

The department of arms is one of the institution's four major divisions and is of comparative recent origin, having been organized in 1946. It has grown considerably in the past decade and now boasts a collection that not only covers the classical periods of Swiss military history but also includes some modern weapons.

The military department has been allotted only three exhibit halls in the entire museum with a total floor space of no more than 3,000 square feet. One hall is quite large, and the remaining two rooms are smaller than the average for most museums. The large hall follows an historical format, with various periods represented by different cases and displays. It is well planned considering the vaulted ceilings, lack of wall space, and other obstacles which had to be overcome in programming the exhibits. Field pieces have been placed in the center of the room, and the few showcases have been pulled away from the walls. A group of armor-clad manikins demonstrating various uses of the pike introduces some variety into the exhibition. The vaulted ceilings prevent the display of many flags, and so a few have been placed in cases. The condition of the armor and weapons is excellent, and the labels are brief, concise, and neatly printed on plexiglass. Only representative artifacts have been included in the exhibits of the main hall. This is a policy followed throughout the entire museum, for only 6 percent of the total inventory is on display.

Of the other two military halls, only one was open to the public in 1958. The second was then being readied for exhibits which would feature modern uniforms and weapons. In contrast to the fine quality of the displays in the main hall, exhibits in the smaller room are extremely crowded. Its cramped size permits only the display of uniforms, hand and shoulder arms, edge weapons, models, and other small items. As there is insufficient space to display the uniforms on manikins, they are mounted against the back of the showcases in a profile or side view arrangement. Hats line the bottom of most cases and musical instruments have been piled on top of some. As a partial solution to the space problem, the small arms have been placed in racks. Hence, in its present state, it resembles to some degree a storage room rather than an exhibit hall.

Although the military department of the Schweizer Landesmuseum is limited in scope, it enjoys a considerable degree of self-sufficiency and a reputation for professional competence. The department is reasonably well staffed and includes a head curator, two assistants, and six technicians. This number of employees permits some degree of specialization within the department as well as a modest research program. Although the staff attempts some extensive study projects, the major portion of its research time is consumed preparing answers to the many questions the public raises about weapons, uniforms, and military history. The museum possesses a good-sized central library, and the military department maintains the bulk of its reference materials there. The central library in Zurich also has many volumes on military history and reference texts on weapons which it keeps on permanent loan to the museum.

The staff of the military department devotes considerable effort to the search for more effective methods of preserving specimens. To that end it has built one of the best-equipped laboratories found in any European museum housing military objects. The laboratory possesses complete facilities for restoring and preserving small arms and edge weapons and for fabricating materials needed in the restoration process. The staff is now directing rather intense inquiry toward finding the best possible method for preserving flags and colors. Various experiments have been conducted, but no final solution has yet been found to this extremely difficult problem.

By definition, the Schweizer Landesmuseum is a national institution. It is administratively controlled by the national government, which also provides its operating funds. Its staff likewise is composed entirely of government employees. The city of Zurich is required to contribute to its support. Maintenance of the building and payment of utilities are the responsibilities which have been placed upon the city government, but these are handled by an annual lump sum appropriation to the museum, and the anticipated cost of these items is included in its budget. The amount which the city of Zurich spends is relatively small to help support a very fine museum whose exhibits attract an average of 120,000 visitors each year.

The military collection on display in the Historisches Museum in Bern fares better spacewise than does the one at the Schweizer Landesmuseum. However, the size of the collection appears to be somewhat smaller, and there is no compelling need to clutter the cases with a mass of objects. The focus of the museum is upon the history of Bern, but since this city is the capital of the nation the museum's point of reference occasionally reaches beyond the confines of the immediate locality. This is certainly true in the case of the military collection, for the objects on exhibit represent equipment in use by the entire Swiss fighting force during particular historical periods. The bulk of the collection is composed of objects from the 16th, 17th, and 18th centuries and features some armor, antique firearms, and edge weapons. Modern arms are not displayed.

The Historisches Museum is well appointed throughout, and the display of military objects is done in excellent taste. Only the finest pieces are shown, and individual cases are well arranged. Specimens are identified and dated, but throughout the entire display there is only a minimum of historical reference material. Thus, the exhibit tends more toward the fine arts than toward a graphic presentation of military history. In scale and form this collection is similar to that contained in other impressive local museums.

Of the three principal general museums in Switzerland, the Musée d'Art et d'Histoire in Geneva displays the largest number of military objects. Essentially, the collection is almost entirely one of 16th- and 17th-century arms and armor, although a few items date from a later period. Thus, the museum features weapons and other personal equipment in use during what was probably the period of Geneva's most vigorous military activity.

The architecture and plan of the Musée d'Art et d'Histoire provide an apt setting for objects of such vintage. The principal military hall is crowned with an ornate beamed ceiling and is divided into two sections by means of several large archways. The many large windows in the building reduce the amount of wall space, and so items tend to be crowded against the small space which is available.

The exhibit techniques employed in the military halls are basically those found in most other European museums of arms and armor. Objects are grouped by category, and a large number of the same type items are displayed. Most of the swords and small arms are in cases in one room, but with few exceptions the suits of armor and associated weapons are not cased. The armor is displayed in several groups, and the pikes and halberds are racked in clusters against the wall. A large number of helmets have been mounted on the walls almost to the level of the ceiling. Hence, they are most difficult to examine in any detail and offer little utility other than providing decorations which are quite appropriate to military halls. All objects are in an excellent state of repair, and the viewer finds much to arrest his interest.

Adjacent to the major military hall is a smaller room which has been constructed in the form of a castle honor hall. This room commemorates military achievements by Geneva arms during the early part of the 17th century. It contains the furnishings of the period and displays only a few characteristic weapons. The simplicity of this room corresponds to that found in a number of other museum halls which display art objects.

The manner in which military collections are displayed in the five excellent general museums discussed above clearly discloses the limitations imposed upon the military curator who must fashion his exhibits within an institution which is concerned with a broad range of subject matter. Confined to a very few rooms, most of which are average in size, he can hope to display little more than small arms, armor, personal equipment, uniforms, and some heraldry. He may be able to bring in a few artillery pieces, but seldom is there the luxury of sufficient space to feature any heavy ordnance or larger objects. He has had to be highly selective in what he does show, and most museums of this type place only their finest objects before the public. Even then some clutter inevitably occurs. The military curator likewise is limited in what he can do with what he does exhibit. He is pretty well precluded from developing any elaborate historical construction and most curators in this situation go little beyond grouping contemporary materials, briefly identifying individual objects, and designating the approximate dates of their use. In spite of such obvious difficulties military curators in these general museums have fashioned exhibits of considerable merit and much interest.

WAFFENSAMMLUNG, KUNSTHISTORISCHES MUSEUM
Vienna, Austria

The Kunsthistorisches Museum in Vienna is one of the most celebrated cultural institutions in western Europe. Many of the treasures assembled by the Hapsburgs have been combined with other objects of historical significance to form its fascinating collections. The Hapsburgs and other prominent leaders of the old Austro-Hungarian Empire were not only patrons of the arts, but also they saw great value in preserving objects of enduring worth as the contents of this museum readily reveal. The Empire's involvement in war has been more properly depicted by the impressive Heeresgeschichtliches Museum, but the personal arms and armor of the Hapsburg emperors and key military leaders of the empire were long held in the Emperor's personal armory.

These have now been placed in the Kunsthistorisches Museum to form the weapons collection or Die Waffensammlung.

A number of the Kunsthistorisches Museum's exhibits, including the Waffensammlung, are housed in the palace buildings in the heart of the city. The weapons collection occupies several long halls and has been arranged chronologically. Basically it is a collection of armor, ranking as one of the foremost of its kind. The other comparable collections in Europe are located at the Tower of London, the gallery housing the Wallace Collection in London, the Real Armería in Madrid, and the Musée de l'Armée in Paris. Each of these features suits of armor and the personal arms of kings, emperors, and famous military commanders, but they also include many other representative pieces whose ownership is quite obscure or cannot be traced. These collections also contain other types of weapons. By contrast, virtually all the objects in the Waffensammlung are fully authenticated as belonging to members of the Hapsburg household or to the chief commanders of the realm. This gives the collection added intrinsic value.

Because much of the armor was designed for the personal use of the Emperor, it was custom-made, and many pieces are now considered among the finest achievements of the armorer's art. A number of ceremonial suits are carved with elaborate and intricate designs, revealing their unknown creators as masters of their craft and artists of rare talent. Some are gold plated, and a number of weapons are jewel encrusted. Also included is the largest quantity of royal tournament armor to be found anywhere. Thus, only part of the armor displayed was actually designed for use in battle.

The preservation of the pieces is particularly excellent. As part of the royal possessions, the armor has been well cared for from the beginning, but some restoration work has been necessary from time to time. With few exceptions, the armor is exposed to the air, and so requires continued maintenance. Repeated cleaning has given these objects a constant gleam, a condition which probably did not exist when they were in actual use.

The Waffensammlung is primarily a fine-arts exhibit. The quality of the objects as well as their historical importance leaves little other display alternative. As a highly specialized collection, it is restricted in scope by its content. It could be the basis for a comprehensive presentation of Austro-Hungarian military history, but much additional supporting material would be needed. Perhaps most wisely, the collection is displayed as an entity, with attention focused upon the objects themselves. Of course, the interest of the viewer is further attracted because these objects are associated with the Hapsburg dynasty.

Organizationally the Waffensammlung is a department within the Kunsthistorisches Museum, and its chief curator has the title of director. Until recently he also supervised the Schatzkammer or royal jewel collection located in an adjacent building. The director is highly esteemed among his colleagues as one of the most competent arms and armor men in Europe. His talents also extend to the general field of military history where he has established an excellent reputation as a scholarly publicist. The Waffensammlung is subject to few changes and requires a minimum of administrative attention. Hence, the director is freed from a number of duties which could seriously restrict the time he finds available for his research projects.

OLDSAKSAMLING
Oslo, Norway

One of Norway's finest museums is located at the University of Oslo. Its exhibits primarily concentrate on ancient and medieval Norwegian history, and so its activities are closely geared to those of the university's department of archeology. The museum emphasizes cultural history, and the weapons displayed in its exhibits are incorporated with other personal objects to demonstrate a variety of human activity.

For a number of years, the department of archeology has conducted studies of the Viking period. In 1867 the first Viking ship was excavated, and two others were found in 1880 and 1904. These ships, together with the many articles found with them, are now preserved in their own hall at Bygdøy, located just outside the city of Oslo. However, the university supervised the work of restoration. Periodically, the university's archeological expeditions have turned up a number of other relics of Viking culture. These have been put on display at the university museum. The total finds now number many objects and include an excellent inventory of weapons used by Viking warriors. The university's collection understandably includes the largest group of Viking weapons in the world and is of vital interest to any student of the implements of warfare.

With few exceptions, the materials of war in use throughout Europe prior to the 14th century have not been preserved. Those of earlier vintage have usually been found through the relentless search of the archeologist. The significance of the Viking weapons thus lies in their association with a period which is still somewhat obscure historically because a paucity of documentary evidence has survived. By the standards of a later date these weapons are somewhat primitive, but they clearly demonstrate that the Vikings had mastered the fundamentals of weapons design and manufacture, at least those which provided the individual warrior with a rather formidable set of personal arms.

Because of its preoccupation with cultural history, the University of Oslo Museum seeks to portray as balanced a picture of Viking civilization as is possible with the materials at hand. The Viking weapons not only provide evidence of military activity but also constitute an extremely useful adjunct to the department of archeology's research undertakings on the Viking period. Although the museum does not give special treatment to military history, the number of weapons discovered as compared with other objects, points up the Vikings' strong warlike tendencies and the degree to which military activity shaped their culture.

For the most part, the arms which have been recovered include swords, daggers, knives, and other metal edge weapons. Other personal equipment of the Viking warrior has only rarely been discovered. Time has dealt rather harshly with many of these objects, for they have often been found in an advanced state of decomposition, and restoration has been most difficult. Hence, it has been necessary to fabricate parts for many weapons to recreate a fairly accurate representation of their original shape. The quality of the restoration work done at the University Museum is indeed excellent, and the staff has employed the very finest modern museum techniques in preparing the entire Viking exhibit. Individual cases are well designed to facilitate close observation and study of their contents. Succinct annotations provide the essential descriptive material.

To a great extent such specialized collections as the Hapsburg armor and the Viking weapons have less flexibility in museological treatment than that which can be accorded a general collection. In the specialized collection attention is naturally focused upon the objects themselves because they are quite unusual or historically significant. The viewer comes to see these celebrated objects, and he would probably raise serious objections if he found them merged with any sizable quantity of instructional material. Hence, collections like the Waffensammlung are often displayed as a unit or as a commemorative exhibit honoring a military hero. The alternatives for displaying such collections as the Viking finds are somewhat greater. They can either be exhibited as a separate entity or merged with other related objects in an historical or cultural context. The latter form of exhibition as employed at the University of Oslo Museum appears to be the more effective. Apart from the unique limitations imposed upon his choice of display methods, the functions, duties, and administrative responsibilities performed by the curator of a specialized military collection are virtually identical with those of his counterpart in the military department of any other general museum.

CASTEL SAN ANGELO
Rome, Italy

Scattered throughout Europe are a number of relatively small military collections, each housed in its own museum. Some began as private collections and were first exhibited in the castle or less pretentious home of their original owner. Eventually many of these passed under some form of public control, and their private character was set aside in favor of their more obvious usefulness as tourist attractions. Other collections were assembled from various sources into a single military museum. Such museums were never intended to attain the size and scope of a national military museum, but a number contain objects which do have national significance, and some have established an excellent professional reputation because of their contributions to the field of military museology. From the standpoint of management, they are self-contained units, and within the limited scope of their activities they confront many of the administrative problems met by the national institution. The weapons collection at the historic

Castel San Angelo in Rome and the Real Armería in Madrid provide useful examples.

The display of arms and armor at the Castel San Angelo is a comparatively recent development for this historic old building located on the bank of the Tiber River near the Vatican. However, the inclusion of a weapons collection is most apropos since throughout its history the Castel San Angelo has functioned as a fortress and sanctuary for the Pope when Rome was under siege.

The origin of the building is somewhat obscure, but its foundations are thought to have been laid before the time of Christ. The storage chambers in its recesses provide convincing evidence that the building must originally have been intended as a strong point in Rome's inner city defenses and designed to keep its defenders self-sufficient for an indefinite period. During the second century A.D., its use was completely transformed, for the superstructure was remodeled to serve as the magnificent tomb of the Emperor Hadrian. How long it remained so is not known, but during one of Rome's periodic upheavals the tomb was despoiled and a part of the building was destroyed. During the medieval period the Castel San Angelo underwent its third and most recent major change. Above the base of what had been the Emperor's tomb there was erected a large turret-type fortress of sufficient height to dominate the surrounding city.

Connecting the Castel San Angelo and the Vatican is an elevated causeway which permitted the Pope speedy access to the protection offered by the fortress whenever his safety was threatened. The Popes have not used the Castel San Angelo for this purpose for many decades, but one of the points of interest in the building today is the fully furnished papal bed chamber and audience room. Indeed, the Castel San Angelo's only major function at present is as a museum.

Apart from the papal objects, the weapons are the only mementos of any significance in the building. These occupy several small rooms on one floor of the upper fortress. The collection is completely compatible with its medieval surroundings, for primarily it consists of 15th-, 16th-, and 17th-century arms and armor, much of which is of Italian manufacture. The rooms containing the weapons are all adjacent and are little more than cubicles opening onto a hallway. The display techniques employed are of the simplest form. Arms and pieces of armor are racked around the walls as if they had been stored in an armory. Others have been placed on tables in the center of the room. None are cased. Each exhibit room is chained off from the public, so that its contents cannot be viewed at very close range. This produces no particular disservice to the average viewer, since none of the rooms are very large. Barring the public likewise keeps the objects from being touched and accords them a certain degree of protection from being damaged or stolen.

There is probably no other weapons collection in Europe housed in a building older than the Castel San Angelo. The architects of the structure obviously did not contemplate its ultimate use as a museum, for the rooms containing the weapons collection certainly were not designed for their present purpose. They are not particularly well lighted and, like the rest of the building, reflect the inevitable changes caused by long and varied use. Modern display techniques could be employed if the rooms underwent some fundamental remodeling and if suitable furnishings were added. Such changes are not planned, at least for the foreseeable future.

The weapons collection is by no means complete, but it is fairly representative of the personal arms and armor used during the centuries covered. The curator of the collection makes no claims for the quality of his objects, but he points out that the collection is the best of its kind to be found in Rome. Also, the viewer receives a reasonably accurate impression of the implements of warfare in use three or four centuries ago. The collection contains no precious items, nor is any effort made to portray any segment of military history. In short, the arms and armor serve primarily as an added attraction for the visitor who comes to explore the interesting labyrinth of the Castel San Angelo's interior. Further, the old fortress appears to be a most logical place to exhibit them, since Rome has no other military museum of major consequence.

The only public employees at the Castel San Angelo are caretakers and guards who are charged with the upkeep and protection of the premises. Their interest in the weapons collection extends no further than seeing that the objects are undisturbed. The curator is unsalaried. He is a local businessman who cares for the objects strictly as a hobby. He possesses a wealth of information and technical competence, but he can devote only his spare time to the collection.

REAL ARMERÍA
Madrid, Spain

In contrast to the weapons displayed at the Castel San Angelo, those found at La Real Armería (the Royal Armory) in Madrid are historically important objects of considerable artistic elegance. This celebrated collection, housed in its own wing of what was once the Royal Palace and is now the residence of General Franco, ranks as one of the major exhibitions of arms and armor in western Europe. In content, it is most closely comparable to the Waffensammlung in the Kunsthistorisches Museum at Vienna.

The personal weapons and armor belonging to Spanish kings of the late 15th and 16th centuries form the nucleus of the Real Armería's collection. Most of these historic items were first preserved in the monarch's private armory and inherited by his successor. The museum's very valuable medieval weapons were brought from the Alcazar at Segovia, the former storehouse of many Spanish royal treasures. Still other pieces of fine armor were obtained from the Sala de los Linajes at Soria where they had been preserved for generations. The most celebrated objects belonged originally to Charles V, Emperor of the Holy Roman Empire, who ruled Spain as Charles I, and to his son, Philip II. Also included are Flemish and Valencian arms belonging to Philip the Handsome, father of Charles V, and some pieces owned by Philip's father, Holy Roman Emperor Maximilian I, which Charles V carefully preserved as part of his personal armory.

Philip II first conceived the idea of placing the family armor together with some battle trophies in a historical museum, and ordered an armory constructed near the Alcazar for this purpose. His successors followed Philip's example and maintained the armory in much its original form. They added their own personal arms, those acquired from their most renowned commanders, and trophies won from Spanish enemies. The armory continued intact until the War of Liberation fought against the French in the days of Napoleon's waning power. During this conflict, many objects disappeared.

In later years the armory underwent two periods of reorganization. The first occurred during the 1840's, and the second took place some 40 years later. The Real Armería opened in its present location and in much its present form in 1893. It suffered some damage during the Spanish Civil War of the late 1930's but was reopened shortly afterward and has continued its steady growth.

The museum has two exhibit floors with only one spacious hall on each. The upper hall is entered from the street level and is approximately two stories in height. The extra height provides an abundance of upper wall space which is effectively used to display a series of ornate tapestries, some colors, and a number of small arms, shields, and other individual pieces of armor. The entire available lower wall space has been fitted with narrow showcases. These contain a number of the Real Armería's more prized and artistic pieces. Individual objects are numbered and can only be identified by reference to the "Illustrated Guide" which the visitor may acquire when he enters.

The upper or principal floor of the museum features the arms and armor of the late 15th and early 16th centuries, together with those of Charles V. The showcases contain a broad range of arms and trophies, the oldest of which is a Visigoth bridle. For the most part the Real Armería's antiquities are to be found in this hall, but so are the most modern objects in its collection—some firearms from the early 19th century.

On the lower floor the arms and armor of Philip II, Philip III, and Philip IV are exhibited together with those belonging to lesser-known Spanish figures. Several suits of armor and weapons used by the kings when they were children provide a special added feature in the lower hall. This floor also includes a sizable collection of crossbows, artillery, swords, banners, saddles, harness, and some oriental arms.

The Real Armería is not principally a museum of military history, although historical references are of necessity included in its descriptive material. Presentation of Spanish military history is primarily the province of the Army and Navy Museums in Madrid. Instead it maintains its character as an armory throughout and focuses the viewer's attention upon the fine objects it displays. As a result, the museum's display techniques are those employed for a fine arts exhibition. For example, the cases contain only a few objects, and each is placed so as to show the intricacies of its design or its most attractive features.

The armor on the upper floor is particularly well displayed. For almost the full length of the hall there are two central rows of figures mounted on horseback and facing each other in what appears to be a formation for individual combat or tournament jousting. Each horse is covered with a colorful cloth caparison and is outfitted with protective armor. The figures atop the horses are completely clad in armor, and all but a very few carry the lance. Behind and between the horses stand a number of complete suits of armor which were once worn in combat, at the tournament or for parades and ceremonies. At one end of the hall is a group of manikins wearing the very elaborate and artistic armor used by Charles V for ceremonial and state occasions. The displays on the lower floor are likewise effective and attractive, but considerably less elaborate.

The quality of the restoration and preservation work performed on the collection is extremely fine. Perhaps the difficulties encountered in keeping the objects in a good state of repair have been diminished somewhat because of the constant excellent care they have received since the days of their actual use. Some items still bear traces of damage they sustained during the Napoleonic wars and the Civil War a century and a quarter later, but sufficient repairs have been made to restore most of their original characteristics. The present state of the collection indicates that the arms and armor technicians at the Real Armería are fully acquainted with modern conservation techniques.

Although the Real Armería adds objects to its collection periodically and maintains an exhibition of very high caliber, its activities are fairly limited. Since this is a museum of arms and armor, the staff feels there is little need to revise its principal displays. Since there is likewise no space available for special exhibits, the museum displays normally remain unchanged except for the occasional addition of new pieces. The small museum staff, headed by a conservator held in very high professional esteem by his European colleagues, is thus largely freed from other than routine upkeep of the collection. Because the Real Armería is primarily concerned with its custodial function, it is managed with a high degree of administrative simplicity. Apart from some increases to its inventory, the museum apparently contemplates no expansion beyond its current operations.

* * *

The smaller, local and provincial military museums have made their own valuable contributions in preserving the weapons and panoply of warfare. They provide adequate settings for specialized and commemorative collections and assist in giving large numbers of people a better acquaintance with the subject of military history and with ancient and modern arms. These museums often provide specialized and detailed treatment of some fascinating military incidents of local significance with which the larger national museum can only deal in a most cursory fashion or must ignore altogether.

Such institutions have an important role, but they can in no wise, singly or in the aggregate, achieve the full potential of a national military museum. It is the larger national institution which seems best able to present the broad sweep of a nation's military history, to display the contributions of the armed forces without becoming immersed in the detailed achievements of local units, to provide greater space for displaying sizable collections and large objects, to conduct reasonably ambitious programs of research, and to possess the facilities needed to carry on many other activities which are required of a modern and dynamic military museum.

EPILOGUE

An intensive study of European military museums produces many vivid and lasting impressions. Separately and collectively these museums provide insights into the many chapters comprising the total history of warfare which has transpired on the European continent. For the most part, their collections of arms, armor, uniforms, and other memorabilia of battle are of very high quality, fairly complete, and generally well preserved. A tour through the museum halls where they are displayed is informative and entertaining both to the casual viewer and the serious student. Although all collections arouse a certain amount of interest, it is usually the weapons and personal items used by celebrated military commanders and sovereigns or items of unusual merit which are long remembered. European military museums are well studded with such objects, and one expects that these would make the strongest impact upon the visitor. But there are other impressions that emerge if the European military museum is seen as an institution which is confronted with and must surmount a variety of organizational and managerial problems. Three of these have been singled out for special mention.

First, the curators and other members of European military museum staffs possess a wealth of experience which can be drawn upon to advantage by either the novice in the field or the expert. One finds among their numbers distinguished scholars in the field of military history whose research efforts have produced a number of standard works in their respective fields. Serving with them are experts whose knowledge covers the broad spectrum of technical problems faced by a museum which preserves and displays military artifacts. Also some museum directors have been coping with the problems of management for many years. Often their experience includes the recovery of collections which were temporarily lost or dispersed during World War II, the reconstruction or refurbishing of their buildings, and sometimes the relocation into new quarters. All these people provide a valuable source for advice covering the full range of problems and activities which are germane to military museums. When requested, this advice is proffered most willingly.

Secondly, modern display techniques are employed in those museums which are financially strong, housed in comparatively new or reconstructed buildings, endowed with ample space, and fortunate in having skilled exhibit technicians on their staffs. But other museums suffering from inadequate buildings and financial and other limitations beyond the control of the museum staffs have largely been precluded from introducing needed or desired changes. Thus, in display methodology they have been unable to keep pace with current developments in the art of exhibition. Observation likewise leads to the conclusion that the United States has made a number of advances in this area and is fully competent to develop its own general expertise.

Finally, as seen through the eyes of an American observer, there is no European military museum which can serve as an exact prototype for a similar national institution which may someday be created in the United States. The armed services of this country have developed their own traditions and have made certain contributions to our national society which are not duplicated in Europe. While these have been depicted in a number of museums scattered throughout the country which display military collections, they have not been given a striking graphic presentation on the scale which exists in some of the national military museums in Europe. How European military museums have dealt with basic museum problems is, of course, of great interest and should be duly studied. However, their solutions to these problems may not always be applicable to the United States. The role of the military in time of peace has also been largely neglected in European museums, and they tend to avoid interpreting significant current problems affecting their nation's security. It is in these fields, as well as in the more comprehensive interpretation of military history, that the United States may be able to develop an entirely unique contribution among military museums.

APPENDIX: SPACE ALLOCATIONS IN EUROPEAN MILITARY MUSEUMS
(All measurements in square feet)

Museum	Interior Exhibit	Exterior Exhibit	Ready Reference and Storage	Library, Study Rooms, & Auditorium	Service Rooms
Imperial War Museum, London	40,000	......	50,000	5,000	......[2]
National Maritime Museum, Greenwich	95,565	924	19,600	10,930	7,200
Musée Royal de l'Armée, Brussels	112,000	......	7,000	17,300	23,000
Leger en Wapenmuseum, Leiden	70,000	6,450	650	830	5,900
Musée de l'Armée, Paris	69,400	......	......[2]	1,300	6,250[3]
Musée de la Marine, Paris	50,570	......	7,500	2,150	14,000
Tøjhusmuseet, Copenhagen	71,000	......	167,850	3,200	8,100
Haermuseet, Oslo[1]	34,500	6,000	32,300	......	......
Armémuseum, Stockholm	45,700	14,100	......	700	2,900
Sjöhistoriska Museum, Stockholm	13,900	17,500	15,000	1,000	7,500
Heeresgeschichtliches Museum, Vienna	58,545	48,266	33,980	5,000	21,000
Museo del Ejercito, Madrid	43,000	8,600	......[2]	750	1,650[3]
Museo Naval, Madrid	21,500	......	......[2]	4,300	3,500

[1] The Haermuseet in Oslo is still in the planning stage, and final space allocations have not been completed.
[2] Not provided.
[3] Does not include workshops and utility service rooms.

INDEX

Akershus Castle, 14, 99, 100, 106
Anderson, Clinton, vii
Armémuseum, Stockholm, viii, 10, 14, 17, 18, 19, 21, 23, 24, 26, 27, 28, 30, 31, 32, 35, 106–117, 159, 203
Arms collections, special, 177–196
Austria, 141, 190

Bavaria, 183
Bayerisches Nationalmuseum, Munich, viii, 11, 183–185
Belgium, 60
Berlin, 130
Brooks, Overton, vii
Brown, John Nicholas, vii
Brussels, 60
Buildings, 2, 12–19

Caird, Sir James, 37, 49, 52
Cannon, Clarence, vii
Carmichael, Leonard, iii, vii
Castel San Angelo, Rome, viii, 14, 15, 18, 21, 194–197
Chateau de Chillon, Montreux, 186
Commemorative function, 7
Copenhagen, 91
Custodial function, 5

Denmark, 91

Educational function, 6
Eisenhower, Dwight D., iii, vii
England, 39, 45, 53, 56
Entertainment function, 7
Exhibition techniques, 2, 19–25

Finance, 4, 36–38
"Friends of the Museum," 26, 32, 37, 71, 78, 79, 80, 87, 90, 97, 116, 124, 127, 129
France, 72, 83

Germany, 130, 181, 183
Greenwich, 45

Haermuseet, Oslo, viii, 14, 29, 31, 32, 33, 35, 99–106, 203

Hamburg, 181
Havernick, Walter, quoted, 6
Heeresgeschichtliches Museum, Vienna, viii, 9, 15, 17, 18, 19, 22, 23, 24, 26, 27, 34, 35, 36, 113, 141–152, 190, 203
Historisches Museum, Bern, 188, 189
Hougen, Bjørn, 102

Imperial War Museum, London, viii, 8, 9, 11, 13, 16, 18, 19, 26, 28, 30, 31, 32, 34, 36, 39–45, 113, 159, 203
Italy, 194

Joyce, Kenyon A., vii, ix

Kunsthistorisches Museum, Vienna, viii, 11, 149, 151, 190–192, 197

Leger en Wapenmuseum (Generaal Hoefer), viii, 11, 14, 18, 19, 23, 24, 29, 31, 32, 34, 35, 66–72, 86, 203
Leiden, 66
Lisbon, 171
London, 39, 53, 56

Madrid, 152, 164, 197
Management and organization, 3, 28–36
May, William E., quoted, 7
McElroy, Neil, vii
Morges arsenal, 186
Munich, 183

Musée d'Art et d'Histoire, Geneva, 189
Musée de la Marine, Paris, viii, 10, 15, 17, 18, 19, 23, 24, 26, 31, 32, 34, 72–83, 113, 117, 167, 203
Musée de l'Armée, Paris, viii, 10, 18, 19, 24, 29, 31, 32, 34, 83–91, 191, 203

Musée Royal de l'Armée et d'Histoire Militaire, Brussels, viii, 9, 15, 16, 18, 19, 26, 29, 31, 33, 34, 60–66, 203
Museo del Ejercito, Madrid, vii, 18, 19, 28, 29, 34, 113, 152–164, 167, 171, 198, 203
Museo Naval, Madrid, viii, 14, 18, 19, 26, 28, 31, 32, 34, 164–170, 171, 198, 203
Museu Militar, Lisbon, viii, 14, 17, 19, 171–177
Museum für Deutsche Geschichte, East Berlin, viii, 130–141
Museum für Hamburgische Geschichte, viii, 181–183

Napoleon Bonaparte, 10, 83, 84, 91, 137, 156
National Maritime Museum, Greenwich, viii, 7, 8, 10, 13, 17, 18, 19, 22, 24, 26, 27, 30, 31, 32, 33, 34, 35, 37, 40, 45–53, 117, 203
Nelson, Lord, 48, 49
Netherlands, 66
Norway, 99, 192

Oldsaksamling, Oslo, viii, 102, 192–194
Organization and management, 3, 28–36
Oslo, 99, 192

Paris, 72, 83
Philosophy, governing military museums, 2, 5–12
President's Committee on the American Armed Forces Museum, iii, vii, ix
Programs and services, 3, 25–28
Portugal, 171

Real Armería, Madrid, viii, 11, 191, 197–200
Rockefeller, Nelson A., vii
Rome, 194
Royal United Service Institution, 40, 56, 57, 58, 60

Royal United Service Museum, London, viii, 13, 18, 23, 28, 56–60

Saltonstall, Leverett, vii
Sandwich, Lord, 49
Schadau Castle, Thun, 186
Schweizer Landesmuseum, Zurich, viii, 24, 186, 187, 188
Seitz, Heribert, quoted, 20, 21, 22
Services and programs, 3, 25–28
Sjohistoriska Museum, Stockholm, viii, 10, 15, 17, 18, 19, 23, 24, 29, 30, 31, 32, 33, 34, 117–130, 166, 203
Smith, H. Alexander, vii
Solothurne armory, 186
Soviet museum philosophy, 130–138
Space allocation, 203
Space problems, 12, 13, 16, 17
Spain, 152, 164, 197
Stockholm, 106, 117
Sweden, 106, 117
Swiss historical museums, 186–190
Switzerland, 196

Tøjhusmuseet, Copenhagen, viii, 11, 14, 17, 18, 19, 21, 23, 24, 26, 28, 29, 33, 34, 86, 91–98, 203
Tower Armouries, London, viii, 11, 13, 18, 21, 30, 40, 53–56, 191
Tower of London, 53–56

University of Oslo Museum, 193

Vienna, 141, 190
Viking weapons, 192, 193, 194
Visitors, 18, 19
Vorys, John, vii
Vostokov, J. I., quoted, 132, 133, 134

Waffensammlung, Kunsthistorisches Museum, Vienna, 190–192, 197
Wallace Collection, London, 191
Warren, Earl, vii
Westrate, J. Lee, iii, ix